MECHANICS OF AERO-STRUCTURES

Mechanics of Aero-structures is a concise textbook for students of aircraft structures, which covers aircraft loads and maneuvers, as well as torsion and bending of single-cell, multi-cell, and open thin-walled structures. Static structural stability, energy methods, and aero-elastic instability are discussed. Numerous examples and exercises are included to enhance students' facility with structural analysis.

This well-illustrated textbook is meant for third- and fourth-year undergraduate students in aerospace and aeronautical engineering programs. The material included can be covered in a one-semester course.

Key features include:

- Torsion and bending of single-cell, multi-cell, and open sections are described in detail.
- Aerodynamic loads, maneuvers, and elementary aero-elastic stability are included.
- The book begins with a description of the aerodynamics loads to motivate the students.
- Includes an in-depth description of energy methods, an essential topic.

Sudhakar Nair has taught aircraft structures for more than 30 years at the Illinois Institute of Technology. He is a Fellow of ASME, an Associate Fellow of AIAA, and a Member of ASEE, Sigma Xi, and Tau Beta Pi. He has authored numerous articles on structural mechanics and applied mathematics and is the author of two previous textbooks: *Introduction to Continuum Mechanics* and *Advanced Topics in Applied Mathematics: For Engineering and the Physical Sciences*. He was Associate Dean for Academic Affairs, Department Chair, and Chair of the Faculty Council at IIT. He received the Barnett award for the best teacher and a special commendation from the AIAA student chapter at IIT.

EASTERN
AIR LINES
FROM THE SMITHSONIAN
NATIONAL AIR AND
SPACE MUSEUM

MECHANICS OF AERO-STRUCTURES

Sudhakar Nair
Illinois Institute of Technology

Shaftesbury Road, Cambridge CB2 8EA, United Kingdom

One Liberty Plaza, 20th Floor, New York, NY 10006, USA

477 Williamstown Road, Port Melbourne, VIC 3207, Australia

314–321, 3rd Floor, Plot 3, Splendor Forum, Jasola District Centre, New Delhi – 110025, India

103 Penang Road, #05–06/07, Visioncrest Commercial, Singapore 238467

Cambridge University Press is part of Cambridge University Press & Assessment, a department of the University of Cambridge.

We share the University's mission to contribute to society through the pursuit of education, learning and research at the highest international levels of excellence.

www.cambridge.org
Information on this title: www.cambridge.org/9781107075771

First published 2015

A catalogue record for this publication is available from the British Library

Library of Congress Cataloging-in-Publication data
Nair, Sudhakar, 1944–
Mechanics of aero-structures / Sudhakar Nair, Illinois Institute of Technology.
pages cm
Includes bibliographical references and index.
ISBN 978-1-107-07577-1 (hardback)
1. Airframes – Design and construction. 2. Aerodynamics. I. Title.
TL671.6.N35 2015
629.132–dc23 2015002304

ISBN 978-1-107-07577-1 Hardback

Additional resources for this publication at www.cambridge.org/nair.

Contents

Preface

This book is intended as a textbook for advanced mechanics of materials for third-year undergraduate students in the area of aerospace engineering and related fields. It is assumed that these students have had a first course in strength of materials and a course in ordinary differential equations in their second year. The material included in this book can be covered in one semester. From my experience in teaching this topic, it has been abundantly clear that students are used to following textbook descriptions topic-by-topic as opposed to following the instructor's presentations, which may be at variance with the chosen text. The large number of excellent textbooks available in physics, calculus, statics, dynamics, and strength of materials has conditioned the students to depend on "one" textbook and "one" notation.

I have followed a logical sequence for introducing students to aero-structures. In Chapter 1, the typical loads expected during a preliminary design of an aircraft are described along with certain essential design considerations such as load factor, proof load, and factor of safety. Also, aerodynamic loads in level flight and under gust conditions are included.

Elements of elasticity from a three-dimensional description to two-dimensional simplification are introduced in Chapter 2. Most students find this a difficult topic. But this is the last chance for them to see the full picture before they go to work or to graduate school to continue structural analysis. Energy methods are explained in Chapter 3 and these are used in the coming chapters wherever they are needed.

Analysis of thin-walled structures under torsion and bending, which is of specific use in aero-structures, is treated in Chapters 4 and 5. Applications to open-, single-, and multiple-cell tubes are emphasized. Shear center (and center of twist) calculations are discussed. Chapter 6 is devoted to elastic stability, including a brief primer on aero-elastic stability.

Chapter 7 considers various failure and yield criteria. Metals as well as epoxy/fiber composites are included. An introduction to fracture mechanics, fatigue, and fatigue crack propagation is also included in this chapter.

I am grateful to my colleague Dr. Roberto Cammino, who provided many suggestions for improvement. My former teacher and thesis advisor, Professor S. Durvasula (Indian Institute of Science, Bangalore), and Professor M. Nambudiripad (National Institute of Technology, Calicut) were inspirational in guiding my career toward elastic structures and I am always indebted to them. I also wish to record my appreciation of my doctoral advisor, the late Eric Reissner, whose name appears when one lists the giants in this field. I also thank the hundreds of aerospace students who took this course with me at the Illinois Institute of Technology and provided me feedback on the material included here.

I thank Peter Gordon and Sara Werden for providing editorial assistance on behalf of Cambridge University Press.

My wife, Celeste, has provided constant encouragement throughout the preparation of the manuscript and I am always thankful to her.

S.N., Chicago

1 Aircraft Structural Components and Loads

The structural components comprising an aircraft may be grouped into three categories: Fuselage, wings, and tail. These three groups interact with each other through mechanical connections and aerodynamic coupling. Their overall shape can be viewed as metal cages wrapped in an aluminum or a composite skin. Fig. 1.1 shows a sketch of a plane with the groups of structures marked. An additional component not considered in this book is the landing gear. The landing gear design is intrinsically connected to shock absorbers and hydraulics.

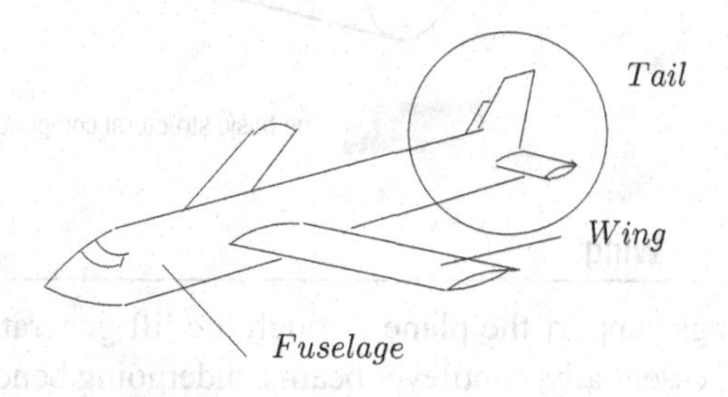

Figure 1.1 The three structural groups of an aircraft.

The details of the tail section are similar to the wing section, but on a smaller scale. That reduces the three groups to two.

From the early days of aviation when wood, cloths, and cables (remember the Wright brothers' biplane) were in use, most of the modern components originated in France and we use many French words, such as *fuselage*, *empennage* (tail section), *longerons* (longitudinal bars), *ailerons* (control surfaces on wings), and so on, to describe structural parts.

1.0.1 Fuselage

Fuselages have circular or oval cross-sections with their dimensions varying along the length of the fuselage. A number of circular (or oval) metal rings called bulkheads are arranged parallel to each other to maintain the circular shape and to form support points for the wings. These bulkheads are connected by longerons (with "Z," "L," or channel shaped cross-sections) to form the basic metal cage. Aluminum sheets cover this cage and form the skin of the plane. The skin is capable of withstanding the pressurization of the cabin, shear stresses due to fuselage torsion, and minor impacts. As a first approximation, the fuselage may be viewed as a beam with concentrated forces acting on it through the points where the wings are connected and through the points where the tail section is joined. Its own weight and aerodynamic forces provide additional distributed loads. Due to asymmetrical loading of the wings and tail arising from control surface movements and wind conditions, there is also a torque acting on the fuselage. Fuselage bending stresses are carried by the longerons. The bulkheads redistribute the forces transmitted from the wings to the fuselage through the longerons. Fig. 1.2 shows the components of a fuselage.

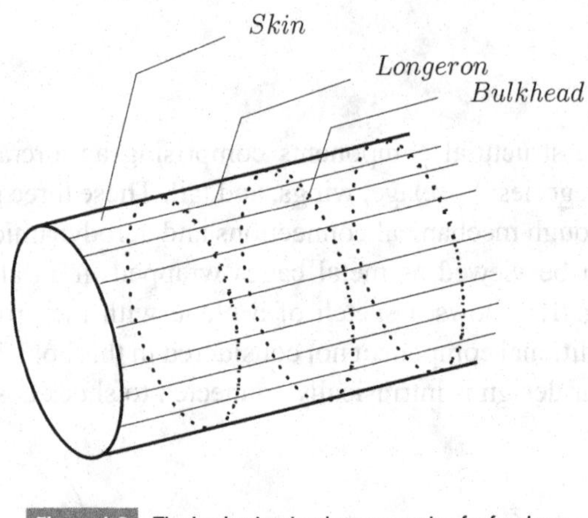

Figure 1.2 The basic structural components of a fuselage.

1.0.2 Wing

Wings support the plane through the lift generated by their airfoil shape. They are, essentially, cantilever beams undergoing bending and torsion. Fig. 1.3 shows the components of a typical wing. Depending on the size of the plane, there

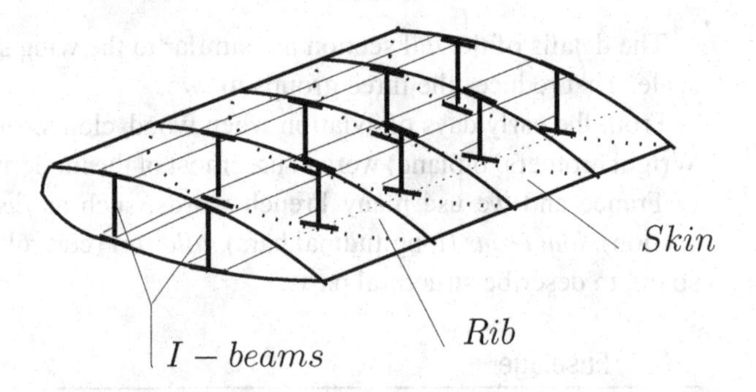

Figure 1.3 The structural components of a wing.

may be one or more I-beams running from the fuselage to the tip of the wing with the needed taper to conform to the airfoil shape. In small planes, instead of the I-beams, metal tubes are used. There are also longerons on the top and bottom surfaces parallel to the beams. The ribs (diaphragms) are used to divide the length of the wing into compartments and to maintain the airfoil shape. The wing is subjected to a distributed aerodynamic loading producing shear force, bending moment, and torque. Concentrated forces may be present if the engines are mounted on the wing. The bending stresses are carried by the beams and the longerons. The skin carries shear stresses due to torsion and the wing volume is used as a fuel tank. Flaps and ailerons are not shown in this sketch.

1.1 Elements of Aerodynamic Forces

A major part of the forces and moments acting on a wing in flight is due to aerodynamics. A specific airfoil shape according to the NACA (forerunner of NASA) classification is usually selected for the wing cross-section. Depending on the design speed of the plane, wings are given dihedral angles, taper, and sweep-back. The aerodynamic forces are computed for various altitudes and airplane speeds. They are then reduced to a reference state known as the "standard level flight" condition. The word "standard" here refers to sea-level values of air density, pressure, and temperature. Sea level is taken as the datum for measuring altitudes. As we know, density, pressure, and temperature strongly depend on the altitude. Fig. 1.4 shows an airfoil at an angle of attack.

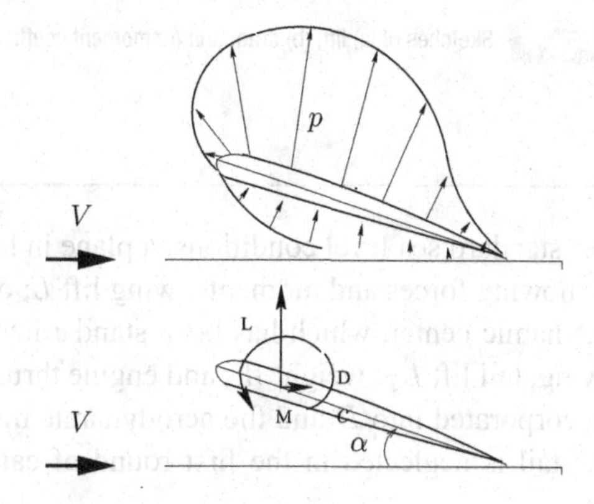

Figure 1.4 An airfoil at an angle of attack α with a pressure distribution and its equivalent forces, L and D, and moment, M, at the quarter chord.

The pressure distribution on the airfoil surfaces is integrated to obtain the wing loading in the form of the lift L, the drag D, and the moment M about the quarter chord of the wing. These can be expressed as

$$L = \frac{1}{2}\rho V^2 S C_L, \quad D = \frac{1}{2}\rho V^2 S C_D, \quad M = \frac{1}{2}\rho V^2 S\bar{c} C_M, \tag{1.1}$$

where ρ is the air density; V is the relative wind speed; S is the plan-form area of the wing; C_L, C_D, C_M are the lift, drag, and the moment coefficients; and $\bar{c}$ is the average chord. We can express the coefficients as

$$C_L = C_{L\alpha}\alpha, \quad C_D = C_0 + C_{D\alpha}\alpha, \quad C_M = C_{M\alpha}\alpha, \tag{1.2}$$

where $C_{L\alpha}$, $C_{D\alpha}$, and $C_{M\alpha}$ are the slopes of the lift, drag, and moment coefficients plotted against the angle of attack α and C_0 is a constant. Generally, within operating ranges of α, $C_{L\alpha}$ and $C_{M\alpha}$ may be taken as constants and $C_{D\alpha}$ may be approximated using a straight line. As the angle of attack, α, increases, C_L

increases and reaches a maximum. The lift suddenly drops beyond the maximum point. This phenomenon is known as stalling. The value of α at stall is called the stall angle of attack. During takeoff and landing, the speed of the plane is low and to obtain sufficient lift to balance the weight, the maximum available lift coefficient is augmented by extended flaps.

The theoretical value for $C_{L\alpha}$, for an airfoil in the idealized shape of a flat plate, is 2π per radian of the angle of attack, which is useful in remembering the approximate upper bound for this constant. Fig. 1.5 shows the typical shapes of the lift, drag, and moment coefficients.

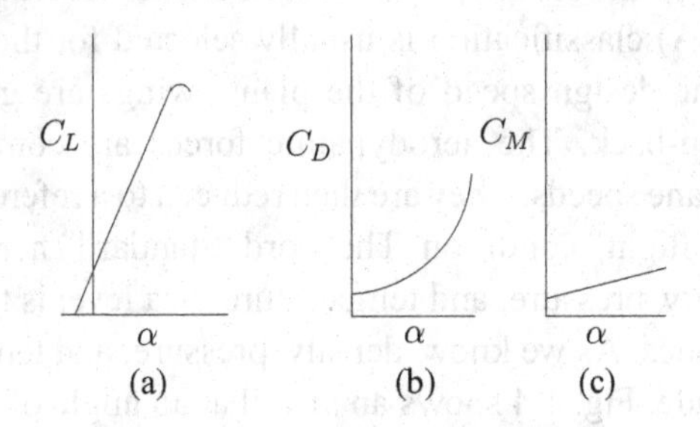

Figure 1.5 Sketches of (a) lift, (b) drag, and (c) moment coefficients as functions of the angle of attack α.

1.2 Level Flight

Under standard sea level conditions, a plane in level flight is in equilibrium with the following forces and moments: wing lift L; drag D; moment M acting at the aerodynamic center, which has been standardized as the quarter chord point of the wing; tail lift L_T; weight W; and engine thrust T. The tail and fuselage drags are incorporated into D and the aerodynamic moment around the quarter chord of the tail is neglected in the first round of calculations. Fig. 1.6 shows these

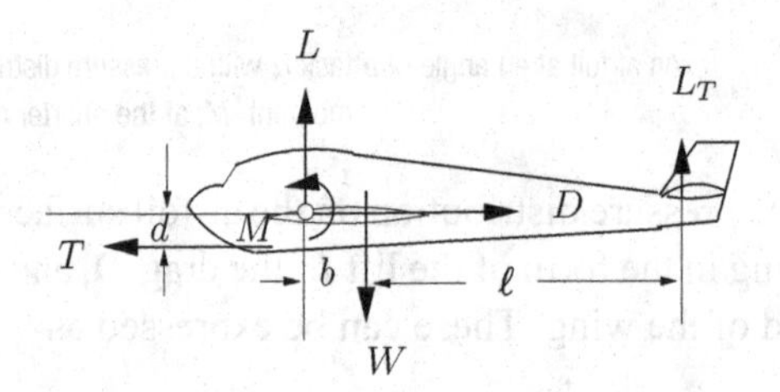

Figure 1.6 Forces and moments on a plane in level flight.

forces for a plane with its engines mounted under the wing. The wing is set at angle of incidence α_i with respect to the fuselage to generate lift during a level flight. For the equilibrium of the plane, we require

$$L + L_T - W = 0, \quad T - D = 0 \tag{1.3}$$

$$Td - M - L_T\ell + Wb = 0. \tag{1.4}$$

The last equation shows the important role of tail lift in balancing the plane.

1.3 Load Factor

Structural members experience amplified loads when a plane undergoes acceleration. The basic loads experienced during level flight are multiplied by a factor known as the load factor to take this amplification into account. We may illustrate

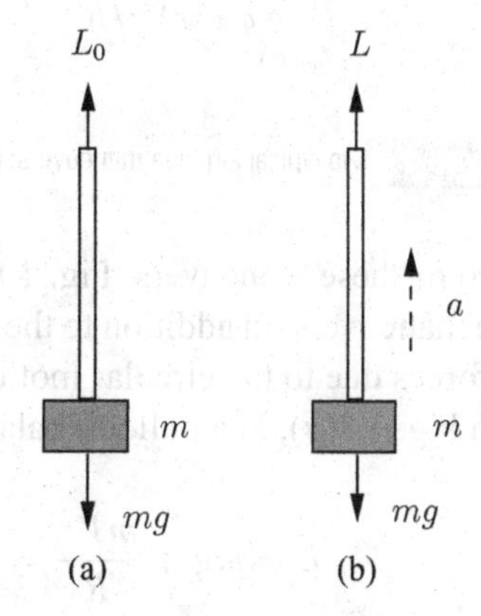

Figure 1.7 Load on a bar under (a) static and (b) dynamic conditions.

this using a simple bar with a cross-sectional area A supporting a mass m. As shown in Fig. 1.7, under static conditions, L_0 balances the weight of the mass, mg, and the stress in the bar is

$$\sigma_0 = \frac{L_0}{A} = \frac{mg}{A} = \frac{W}{A}. \tag{1.5}$$

If we give a vertical acceleration a to this system, for dynamic equilibrium, Newton's second law requires

$$L = m(g + a), \quad \sigma = \frac{mg}{A}\left(1 + \frac{a}{g}\right). \tag{1.6}$$

Then the load factor n is obtained as

$$n = \frac{\sigma}{\sigma_0} = \frac{L}{W} = 1 + \frac{a}{g}. \tag{1.7}$$

Of course, under static conditions $n = 1$.

We use n as a load factor for accelerations perpendicular to the fuselage. There are cases where the plane may be pitching and the acceleration is centrifugal (parallel to the fuselage). The load factor in that case is distinguished using the notation n_x.

1.4 Maneuvers

Even if the plane is not designed for aerobatics, there are certain maneuvers all planes have to undergo. Pullout (or pull-up) from a dive (or descent) and banked

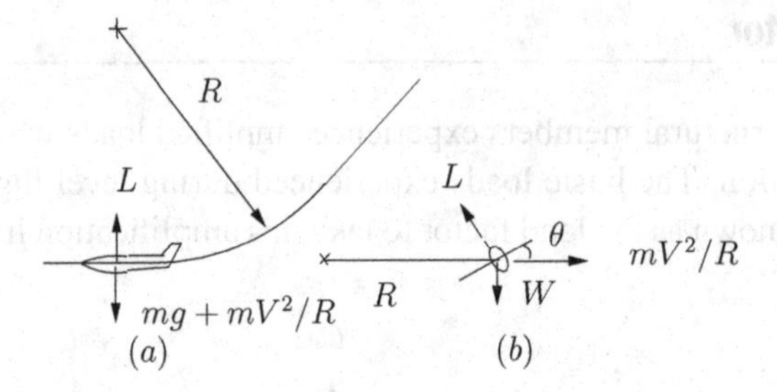

Figure 1.8 Two typical airplane maneuvers: (a) Pullout from a dive and (b) banked turn.

turns are two of these maneuvers. Fig. 1.8 shows the forces acting on the plane during these maneuvers. In addition to the weight and the lift forces, we also have centripetal forces due to the circular motions.

As shown Fig. 1.8(a), in a pullout, balance of forces gives

$$L = mg + \frac{mV^2}{R} = mg\left(1 + \frac{V^2}{gR}\right). \tag{1.8}$$

The load factor for this case is

$$n = \frac{L}{W} = 1 + \frac{V^2}{gR}. \tag{1.9}$$

Fig. 1.8(b) shows a banked turn, with a bank angle θ. To balance the forces,

$$L^2 = W^2 + (mV^2/R)^2, \quad L = W\sqrt{1 + \frac{V^4}{g^2R^2}}. \tag{1.10}$$

Then,

$$n = \sqrt{1 + \frac{V^4}{g^2R^2}}. \tag{1.11}$$

We also have

$$L\cos\theta = W, \quad n = \frac{L}{W} = \frac{1}{\cos\theta}. \tag{1.12}$$

For example, when $\theta = 60°$, we get $n = 2$.

The load factor n is also known as the *g-factor* in everyday language.

1.5 Gust Load

Consider a plane in level flight with a speed V hit with an upward gust with speed v. Some times the gust speed builds up slowly, and we use the term "graded gust" to describe that situation. The situation we have is a "sharp-edged" gust.

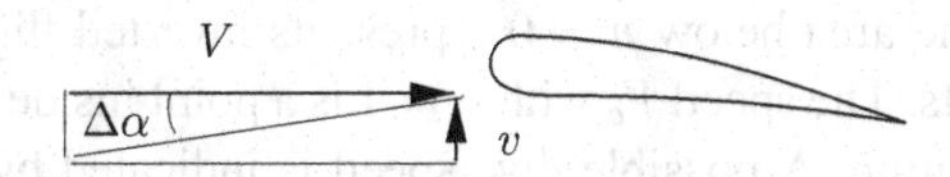

Figure 1.9 Effective increase in angle of attack due to a vertical gust.

As shown in Fig. 1.9, the direction of the net airflow has changed and the effective angle of attack of the airfoil has increased. If $v \ll V$, the change in the angle of attack is

$$\Delta\alpha = \frac{v}{V}. \tag{1.13}$$

Using

$$L = \frac{1}{2}\rho V^2 S C_{L\alpha}\alpha, \tag{1.14}$$

the change in the lift is

$$\Delta L = \frac{1}{2}\rho V^2 S C_{L\alpha}\Delta\alpha. \tag{1.15}$$

If $L/W = n$, the load factor will change by the amount

$$\Delta n = \frac{1}{2}\frac{\rho V^2 S C_{L\alpha}}{W}\frac{v}{V}. \tag{1.16}$$

Of course, this change would result in added loads on the structures.

1.6 *V–n* Diagram

A diagram with the speed V of the plane on one axis and the load factor n on the other axis is known as the V–n diagram. This diagram shows a restricted corridor in which a plane is designed to operate. The maximum values of n for normal flight and for inverted flights are given as part of the design specification. Military aircraft require much larger values of n compared with civil aircraft. At takeoff, with the flaps extended, the lift coefficient reaches a maximum value and there is a minimum speed known as the takeoff speed to lift the plane from the ground. At level flight, the load factor is unity. Maximum values of n may occur during a pullout from a dive or during a banked turn. Fig. 1.10 shows a sketch of the V–n diagram.

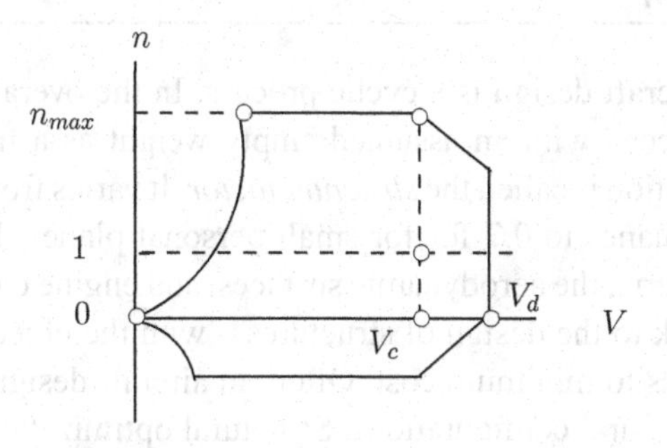

Figure 1.10 Typical shape of the V–n diagram.

The area below $n = 0$ represents inverted flights and the area above normal flights. The speed V_c with $n = 1$ is a point inside the diagram representing cruise condition. A possible dive speed is indicated by V_d. For a plane to lift off, the flaps are deflected to attain the maximum lift curve slope C_{Lmax} and the speed is increased to V_{stall}. Balancing the lift and weight, we find

$$n = 1 = \frac{1}{2}\frac{\rho S C_{Lmax}}{W} V_{stall}^2. \tag{1.17}$$

When the speed is higher than V_{stall}, n increases with the speed V parabolically as

$$n = \frac{1}{2}\frac{\rho S C_{Lmax}}{W} V^2. \tag{1.18}$$

For a semi-aerobatic plane, the load factor is bounded as $-1.8 < n < 4.5$ and for a fully aerobatic plane, $-3 < n < 6$. Generally, these bounds are specified by the customers.

1.7 Proof Load and Ultimate Load

With the expected maximum load factor n_{max}, we have to use a factor safety in the actual strength of any designed component. This factor is considered in two steps. In the first step, we multiply n_{max} by a factor of 1.25 to obtain the proof load factor. The meaning of the term "proof load" is that an actual manufactured structural component can be subjected to the proof load in a laboratory without it sustaining any permanent damage. In the second step, we multiply n_{max} by a factor of 1.5 to get the ultimate load. The factors 1.25 and 1.5 are the recommended values from the cumulative experience of aircraft design over the years. A component is expected to survive up to the ultimate load. Compared to mechanical and civil engineering structures, the demand for weight reduction in aero-structures limits the margin of safety in our field. In addition, we also have to take into account the statistical variability in the mechanical properties of materials. Modern manufacturing processes endeavor to minimize property variations by controlling the chemical compositions of the materials, the heat treatment, and the machining.

1.8 Optimization

Aircraft design is a cyclic process. In the overall takeoff weight calculation, we proceed with an assumed empty weight as a fraction of the total weight. This fraction is called the *structure factor*. It varies from about 0.5 for large commercial airplanes to 0.7 for for small personal planes. From this step, the performance criteria, the aerodynamic surfaces, and engine capacity are obtained. Then, we go back to the design of structures – with the objective of minimum weight, which leads to minimum cost. Often, in aircraft design, we face multiple solutions for structural configurations. Structural optimization to attain minimum weight is a computationally intensive process which is worthwhile considering the reduction in weight and the savings in fuel consumption. Material selection is affected by the cost and the reduction in weight. Detailed design also uses extensive finite element computations of stresses and displacements to obtain accurate values for expected stresses.

FURTHER READING

Abbott, I. and Von Doenhoff, A., *Theory of Wing Sections*, Dover (1949).
Arora, J. S., *Introduction to Optimum Design*, McGraw-Hill (1989).
Corke, T. C., *Design of Aircraft*, Prentice Hall (2003).
Dassault Systemes, ABAQUS Inc., Finite Element Software for Stress Analysis.
Haftka, R. T. and Kamat, M. P., *Elements of Structural Optimization*, Martinus Nijhoff (1985).
Nicolai, L., *Fundamentals of Aircraft Design*, University of Dayton (1975).
Perkins, C. D. and Hage, R. E., *Airplane Performance Stability and Control*, John Wiley (1949).
Zienkiewicz, O. C., *The Finite Element Method*, McGraw-Hill (1979).

EXERCISES

1.1 A plane is modeled as a uniform beam of length 15 m. The fuselage weighs 3 MN per meter. Measuring x from the nose, the weight of the two wings and engines, 30 MN, is applied at $x = 7$ m. The wing lift acts at $x = 6$ m and the tail lift acts at $x = 14$ m. Balancing the plane, obtain the values of the wing and tail lifts. Draw the bending moment and shear force diagrams for level flight. If this plane experiences an upward acceleration of 3 g, what are the maximum bending moment and shear force?

1.2 A vertical bar of length 2 m in an airplane carries a mass of 1000 kg at its lower end. The anticipated load factor for this structure is 3.5. What are the limit load, the proof load, and the ultimate load for this bar based on the factors given in Section 1.7? Design the minimum diameter of the bar if the material fails at 450 MPa. Neglect the weight of the bar in the first attempt. If the weight is included, assuming the bar is made of an aluminum alloy with density 2.78 g/cm^3, what is the new diameter?

1.3 A wing is modeled as a cantilever beam generating a total lift of 32 MN. It has a length of 8 m. An engine weighing 10 MN is mounted at the mid-point. Draw the shear force and bending moment diagrams and obtain their maximum values.

1.4 A plane of mass 3000 kg is in level flight at a speed of 300 km/hr with the wings set at an angle of incidence of 3°. If a sharp-edged vertical gust hits the plane with a speed of 30 km/hr, what is the load factor for the plane?

1.5 A wing has a chord of 3 m and it is set at an incident angle of 4° at level flight at sea level. The lift curve slope for the wing, $C_{L\alpha}$, is 4 per radian. Obtain the wing span for a rectangular wing to support a plane of mass 3000 kg flying at a speed of 300 km/hr. The air density at sea level is 1.226 kg/m^3. Neglect the lift contribution from the tail.

1.6 A plane diving at an angle of 15° with the horizon with a speed of 200 km/hr is leveled by pulling up a curve of radius R. If the maximum load factor allowed is 3, compute the minimum value of R.

1.7 A plane weighing 30 MN is executing a banked turn at an angle of 40°. Compute the load factor for this maneuver if there is no slipping.

2 Elements of Elasticity

2.1 Traction Vector

2.1.1 External and Internal Forces

We consider a three-dimensional elastic body in static equilibrium with external forces $\boldsymbol{P}_1$, $\boldsymbol{P}_2$, and so on. Equilibrium implies that these forces and their moments add up to zero.

As shown in Fig. 2.1, if we separate this body into two parts by an arbitrary cut, the two parts have to be in equilibrium. This reveals internal distributed forces on the area A at the cut to balance the external forces. A close-up look at a small triangular area ΔA shows a traction vector $\boldsymbol{T}^n$, where the superscript n indicates the unit normal $\boldsymbol{n}$ to the cut. The dimension of $\boldsymbol{T}^n$ is force per unit area. The

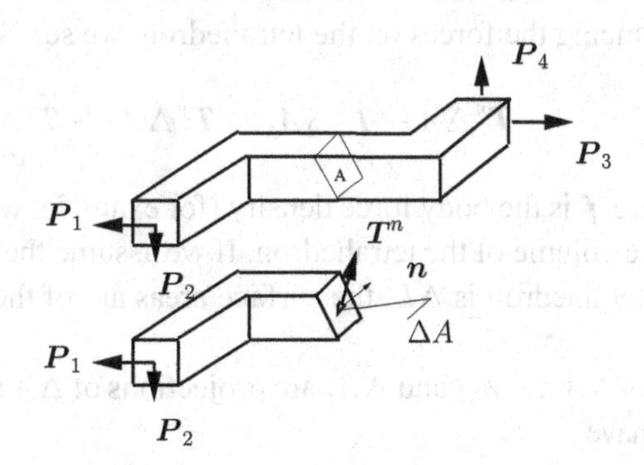

Figure 2.1 Traction vector at a point on a section.

traction vector will vary from point to point on the section A. Also, the cut we have made produces two surfaces: one with normal $\boldsymbol{n}$ and the other with normal $-\boldsymbol{n}$. If $\boldsymbol{T}^n$ is the traction vector on the surface with normal $\boldsymbol{n}$, the surface facing it has the normal $-\boldsymbol{n}$ and the traction vector $-\boldsymbol{T}^n$. To describe the rules governing the variation of $\boldsymbol{T}^n$ across the cut we need a coordinated system. At this stage, all we require is a locally orthogonal system. This may be a Cartesian system with x, y, and z axes, a cylindrical system with r, θ, and z, or a spherical system with R, θ, and ϕ. For the sake of clarity we will use the Cartesian system.

By drawing parallels to the coordinate axes through the corners of our triangular area ΔA, we create a tetrahedron with its inclined facet being ΔA with unit normal

$$\boldsymbol{n} = n_x \boldsymbol{i} + n_y \boldsymbol{j} + n_z \boldsymbol{k}. \tag{2.1}$$

The normals to the three facets that are parallel to the coordinate planes are: $-\boldsymbol{i}$, $-\boldsymbol{j}$, and $-\boldsymbol{k}$. We refer to these four normals as outward normals. As shown in Fig. 2.2, the area of the facet normal to the coordinate direction x is denoted by ΔA_x, and so on.

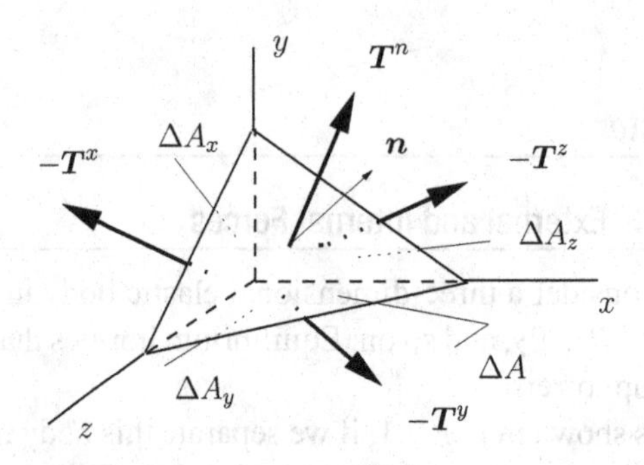

Figure 2.2 Traction vectors on coordinate facets.

The traction vectors on the three coordinate facets are denoted by $-\boldsymbol{T}^x$, $-\boldsymbol{T}^y$, and $-\boldsymbol{T}^z$. The minus signs are needed as the normals are along $-\boldsymbol{i}$, and so on. Balancing the forces on the tetrahedron, we see

$$\boldsymbol{T}^n \Delta A - \boldsymbol{T}^x \Delta A_x - \boldsymbol{T}^y \Delta A_y - \boldsymbol{T}^z \Delta A_z + \boldsymbol{f} \Delta V = \boldsymbol{0}, \tag{2.2}$$

where $\boldsymbol{f}$ is the body force density (for example, weight per unit volume) and ΔV is the volume of the tetrahedron. If we assume the maximum length of an edge of the tetrahedron is ΔL, the surface areas are of the order of ΔL^2 and the volume ΔL^3.

As ΔA_x, ΔA_y, and ΔA_z are projections of ΔA along the coordinate directions, we have

$$\lim_{\Delta L \to 0} \left\{ \frac{\Delta A_x}{\Delta A}, \frac{\Delta A_y}{\Delta A}, \frac{\Delta A_z}{\Delta A} \right\} = \{n_x, n_y, n_z\}, \qquad \lim_{\Delta L \to 0} \frac{\Delta V}{\Delta A} = 0. \tag{2.3}$$

Dividing Eq. (2.2) by ΔA and taking the limit, we get

$$\boldsymbol{T}^n = \boldsymbol{T}^x n_x + \boldsymbol{T}^y n_y + \boldsymbol{T}^z n_z. \tag{2.4}$$

The implication of this fundamental relation is that if we know the three traction vectors, $\boldsymbol{T}^x$, $\boldsymbol{T}^y$, and $\boldsymbol{T}^z$, on the three coordinate facets, we can compute the traction vector, $\boldsymbol{T}^n$, on any inclined plane with normal $\boldsymbol{n}$.

It is conventional to refer to the coordinate planes as xy-plane, yz-plane, and zx-plane. Using the direction of the normal to these planes, we may also refer to them as z-plane, x-plane, and y-plane, respectively.

2.2 Stress Components

Let us look at the traction vector $\boldsymbol{T}^x$ acting on the x-facet (now facing in the $\boldsymbol{i}$ direction). As shown in Fig. 2.3, we may express it using its components as

$$\boldsymbol{T}^x = \sigma_{xx}\boldsymbol{i} + \sigma_{xy}\boldsymbol{j} + \sigma_{xz}\boldsymbol{k}, \tag{2.5}$$

where σ_{xx} is the normal stress and σ_{xy} and σ_{xz} are shear stresses. The first subscript indicates the plane and the second subscript the direction of the force.

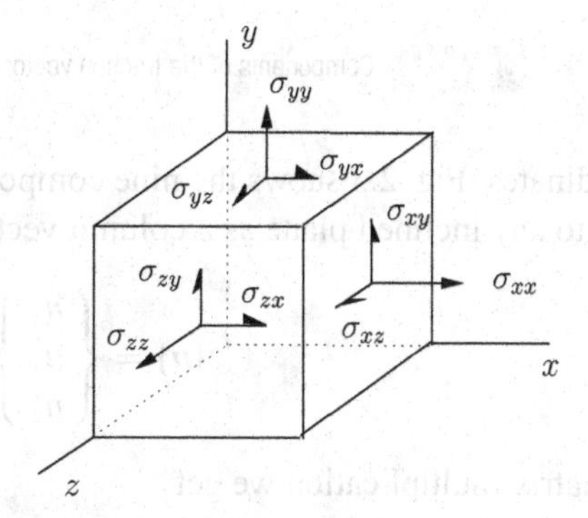

Figure 2.3 Stress components on the three coordinate planes.

Similarly, the other two traction vectors can be resolved as

$$\boldsymbol{T}^y = \sigma_{yx}\boldsymbol{i} + \sigma_{yy}\boldsymbol{j} + \sigma_{yz}\boldsymbol{k} \tag{2.6}$$

$$\boldsymbol{T}^z = \sigma_{zx}\boldsymbol{i} + \sigma_{zy}\boldsymbol{j} + \sigma_{zz}\boldsymbol{k}. \tag{2.7}$$

In terms of the stress components the traction vector $\boldsymbol{T}^n$ can be written as

$$\begin{aligned} \boldsymbol{T}^n = & (\sigma_{xx}\boldsymbol{i} + \sigma_{xy}\boldsymbol{j} + \sigma_{xz}\boldsymbol{k})n_x \\ & + (\sigma_{yx}\boldsymbol{i} + \sigma_{yy}\boldsymbol{j} + \sigma_{yz}\boldsymbol{k})n_y \\ & + (\sigma_{zx}\boldsymbol{i} + \sigma_{zy}\boldsymbol{j} + \sigma_{zz}\boldsymbol{k})n_z. \end{aligned} \tag{2.8}$$

Collectively, we need nine components of stress to create three traction vectors on the coordinate planes; they in turn can give us the traction vector on any inclined plane. We started with a cut and a point on the cut to introduce the nine components of the stress. There are these nine components at every point of our three dimensional body. These nine components form the stress matrix or the stress tensor,

$$[\sigma] = \begin{bmatrix} \sigma_{xx} & \sigma_{xy} & \sigma_{xz} \\ \sigma_{yx} & \sigma_{yy} & \sigma_{yz} \\ \sigma_{zx} & \sigma_{zy} & \sigma_{zz} \end{bmatrix}. \tag{2.9}$$

The rows of the matrix indicate the planes and the columns the directions of the forces. It is called a tensor based on how the components transform under rotation

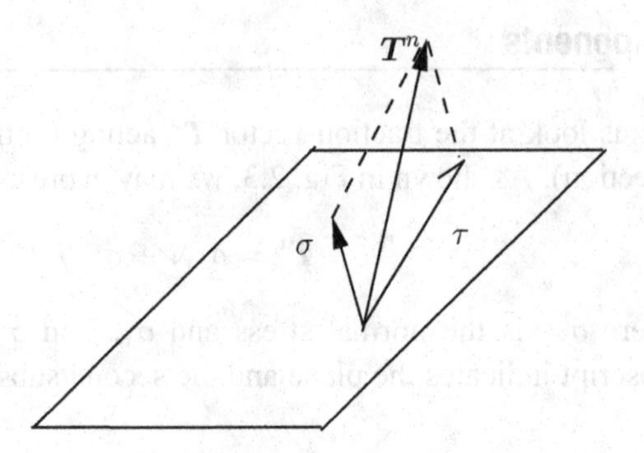

Figure 2.4 Components of the traction vector normal and parallel to the plane.

of coordinates. Fig. 2.3 shows the nine components of stress. If we write the unit normal to any inclined plane as a column vector

$$\{\boldsymbol{n}\} = \begin{Bmatrix} n_x \\ n_y \\ n_z \end{Bmatrix}, \tag{2.10}$$

using matrix multiplication we get

$$\begin{Bmatrix} T_x^n \\ T_y^n \\ T_z^n \end{Bmatrix} = \begin{bmatrix} \sigma_{xx} & \sigma_{yx} & \sigma_{zx} \\ \sigma_{xy} & \sigma_{yy} & \sigma_{zy} \\ \sigma_{xz} & \sigma_{yz} & \sigma_{zz} \end{bmatrix} \begin{Bmatrix} n_x \\ n_y \\ n_z \end{Bmatrix}, \tag{2.11}$$

where the column matrix on the left hand side represents the components of $\boldsymbol{T}^n$.

With $\boldsymbol{n}$ being the unit normal to the inclined plane, please note that, in general, the traction vector is not aligned with the normal. The component of $\boldsymbol{T}^n$ along $\boldsymbol{n}$ is the normal stress σ on this plane. This can be expressed in multiple ways as

$$\sigma = \boldsymbol{n} \bullet \boldsymbol{T}^n = n_x T_x^n + n_y T_y^n + n_z T_z^n = \{\boldsymbol{n}\}^T [\sigma]^T \{\boldsymbol{n}\}, \tag{2.12}$$

$$= \sigma_{xx} n_x^2 + \sigma_{yy} n_y^2 + \sigma_{zz} n_z^2$$
$$+ (\sigma_{xy} + \sigma_{yx}) n_x n_y + (\sigma_{yz} + \sigma_{zy}) n_y n_z + (\sigma_{zx} + \sigma_{xz}) n_z n_x, \tag{2.13}$$

where the superscript T indicates transpose of the matrix. The component of $\boldsymbol{T}^n$ parallel to the plane is the shear stress

$$\boldsymbol{\tau} = \boldsymbol{T}^n - \sigma \boldsymbol{n}. \tag{2.14}$$

This is shown in Fig. 2.4.

2.2.1 Example: Traction Vector on a Plane

Consider the stress matrix at a point (the units may be MPa)

$$[\sigma] = \begin{bmatrix} 3 & 3 & 0 \\ 3 & 4 & 2 \\ 0 & 2 & 5 \end{bmatrix}$$

and a plane with the unit normal

$$\boldsymbol{n} = \frac{1}{3}(2\boldsymbol{i} + 2\boldsymbol{j} + \boldsymbol{k}).$$

Here, the matrix is symmetric and its transpose is the same as the matrix. Then

$$\begin{Bmatrix} T_x^n \\ T_y^n \\ T_z^n \end{Bmatrix} = \frac{1}{3}\begin{bmatrix} 3 & 3 & 0 \\ 3 & 4 & 2 \\ 0 & 2 & 5 \end{bmatrix}\begin{Bmatrix} 2 \\ 2 \\ 1 \end{Bmatrix} = \frac{1}{3}\begin{Bmatrix} 12 \\ 16 \\ 9 \end{Bmatrix}.$$

$$\boldsymbol{T}^n = \frac{1}{3}(12\boldsymbol{i} + 16\boldsymbol{j} + 9\boldsymbol{k}).$$

The normal stress σ on this plane is the normal component of $\boldsymbol{T}^n$,

$$\sigma = \boldsymbol{n} \bullet \boldsymbol{T}^n = \frac{65}{9} = 7.222.$$

The shear stress on this plane is

$$\boldsymbol{\tau} = \boldsymbol{T}^n - \sigma\boldsymbol{n} = \frac{1}{3}(12\boldsymbol{i} + 16\boldsymbol{j} + 9\boldsymbol{k}) - \frac{65}{27}(2\boldsymbol{i} + 2\boldsymbol{j} + \boldsymbol{k})$$

$$= \frac{1}{27}(-22\boldsymbol{i} + 14\boldsymbol{j} + 16\boldsymbol{k}),$$

which has a magnitude

$$\tau = 1.133.$$

2.3 Equilibrium Equations

2.3.1 Force Equilibrium

The equilibrium equations are three partial differential equations describing the changes in the stress components from point to point. Consider the parallelepiped shown in Fig. 2.5, which has side lengths of Δx, Δy, and Δz along x, y, and z directions, respectively.

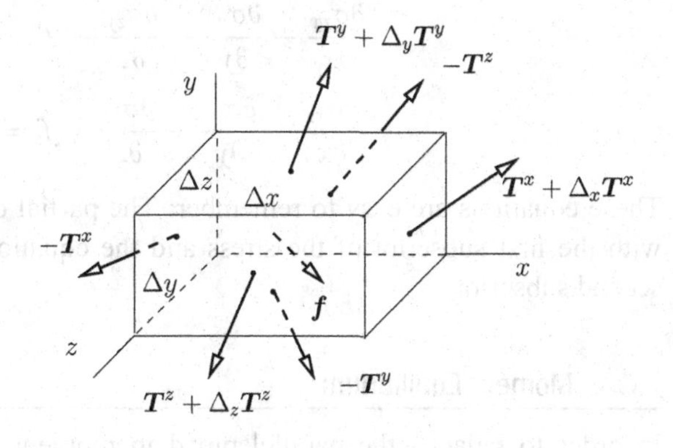

Figure 2.5 Forces on the facets of a parallelepiped.

The origin of the axes is located at the point (x, y, z). If the negative-facing x-facet has the traction vector $-\boldsymbol{T}^x$, the positive-facing facet, which is Δx away, will have the traction vector $\boldsymbol{T}^x + \Delta_x \boldsymbol{T}^x$. By $\Delta_x \boldsymbol{T}^x$, we mean a small change in $\boldsymbol{T}^x(x, y, z)$ when x changes to $x + \Delta x$ while y and z are kept constant. This can also be interpreted as a small change along the x-direction. In fact,

$$\lim_{\Delta x \to 0} \frac{\Delta_x \boldsymbol{T}^x}{\Delta x} = \frac{\partial \boldsymbol{T}^x}{\partial x}. \tag{2.15}$$

Let us use the notation for areas of the facets and the volume of the parallelepiped,

$$\Delta A_x = \Delta y \Delta z, \quad \Delta A_y = \Delta z \Delta x, \quad \Delta A_z = \Delta x \Delta y, \quad \Delta V = \Delta x \Delta y \Delta z. \tag{2.16}$$

As the areas of the positive and negative x-facets are the same, the net force from these two facets is $\Delta_x \boldsymbol{T}^x \Delta A_x$. Including the pairs of traction vectors on the y- and the z-facets and the body force $\boldsymbol{f} \Delta V$ (body forces are distributed forces per unit volume of the body due to the weight of the body or inertial forces arising from accelerations), we balance the forces to get

$$\Delta_x \boldsymbol{T}^x \Delta A_x + \Delta_y \boldsymbol{T}^y \Delta A_y + \Delta_z \boldsymbol{T}^z \Delta A_z + \boldsymbol{f} \Delta V = \boldsymbol{0}.$$

Dividing by ΔV and simplifying, we have

$$\frac{\Delta_x \boldsymbol{T}^x}{\Delta x} + \frac{\Delta_y \boldsymbol{T}^y}{\Delta y} + \frac{\Delta_z \boldsymbol{T}^z}{\Delta z} + \boldsymbol{f} = \boldsymbol{0}.$$

Taking the limit, we get

$$\frac{\partial \boldsymbol{T}^x}{\partial x} + \frac{\partial \boldsymbol{T}^y}{\partial y} + \frac{\partial \boldsymbol{T}^z}{\partial z} + \boldsymbol{f} = \boldsymbol{0}. \tag{2.17}$$

Using

$$\boldsymbol{f} = f_x \boldsymbol{i} + f_y \boldsymbol{j} + f_z \boldsymbol{k}, \tag{2.18}$$

and the expressions for the traction vectors from Eqs. (2.5–2.7), the above vector equation can be written as

$$\begin{aligned} &\frac{\partial \sigma_{xx}}{\partial x} + \frac{\partial \sigma_{yx}}{\partial y} + \frac{\partial \sigma_{zx}}{\partial z} + f_x = 0, \\ &\frac{\partial \sigma_{xy}}{\partial x} + \frac{\partial \sigma_{yy}}{\partial y} + \frac{\partial \sigma_{zy}}{\partial z} + f_y = 0, \\ &\frac{\partial \sigma_{xz}}{\partial x} + \frac{\partial \sigma_{yz}}{\partial y} + \frac{\partial \sigma_{zz}}{\partial z} + f_z = 0. \end{aligned} \tag{2.19}$$

These equations are easy to remember: The partial differential operator agrees with the first subscript of the stress and the equation number agrees with the second subscript.

2.3.2 Moment Equilibrium

In order to balance the parallelepiped in moment, we may use the in-plane forces creating a moment about the z-axis. As shown in Fig. 2.6, in computing

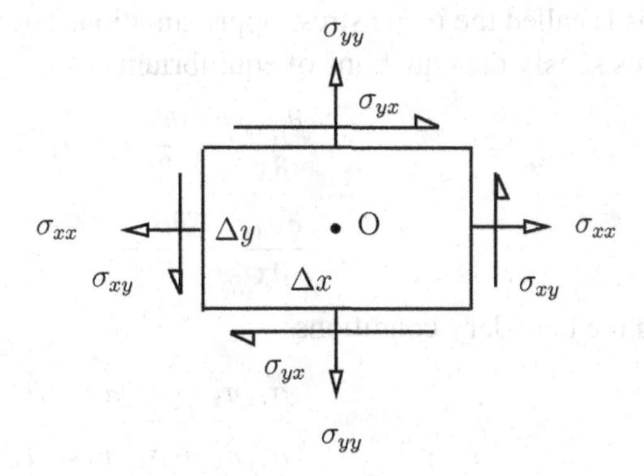

Figure 2.6 Moment balance in the z-plane.

the moment about O, the forces normal to the facets have no effect; the shear stresses give

$$2\sigma_{xy}\Delta y\Delta z\Delta x/2 - 2\sigma_{yx}\Delta x\Delta z\Delta y/2 = 0.$$

This and similar equations about the other two axes yield

$$\sigma_{xy} = \sigma_{yx}, \quad \sigma_{yz} = \sigma_{zx}, \quad \sigma_{zx} = \sigma_{xz}. \tag{2.20}$$

Thus, the stress matrix is symmetric. Up to this point, we were careful to distinguish the off-diagonal elements of the stress matrix, but from now on we need not do this. The number of stress components reduces to six when symmetry is applied.

2.3.3 Traction Boundary Conditions

On certain parts of the boundary of the elastic body we may apply distributed forces. This is equivalent to prescribing the traction vector $\boldsymbol{T}^n$ at every point on those parts of the body. In component form, we prescribe

$$\sigma_{xx}n_x + \sigma_{xy}n_y + \sigma_{xz}n_z = \bar{T}^n_x, \tag{2.21}$$

$$\sigma_{xy}n_x + \sigma_{yy}n_y + \sigma_{yz}n_z = \bar{T}^n_y, \tag{2.22}$$

$$\sigma_{xz}n_x + \sigma_{yz}n_y + \sigma_{zz}n_z = \bar{T}^n_z, \tag{2.23}$$

where the bar over T indicates that these are known (given) functions and $\boldsymbol{n}$ is the unit normal to the boundary.

2.4 Plane Stress

If the elastic body is thin in one direction compared with the other two directions and if there is no force applied in that direction, we can orient our coordinates to have the z-axis perpendicular to the thin plate and use the approximation

$$\sigma_{zz} = 0, \quad \sigma_{xz} = 0, \quad \sigma_{yz} = 0. \tag{2.24}$$

This is called the plane stress approximation. The remaining three stress components satisfy the equations of equilibrium

$$\frac{\partial \sigma_{xx}}{\partial x} + \frac{\partial \sigma_{xy}}{\partial y} + f_x = 0, \tag{2.25}$$

$$\frac{\partial \sigma_{xy}}{\partial x} + \frac{\partial \sigma_{yy}}{\partial y} + f_y = 0, \tag{2.26}$$

and the boundary conditions

$$\sigma_{xx} n_x + \sigma_{xy} n_y = \bar{T}_x^n, \tag{2.27}$$

$$\sigma_{xy} n_x + \sigma_{yy} n_y = \bar{T}_y^n. \tag{2.28}$$

As most of the aero-structures are thin and the in-plane stresses are large compared to the applied load perpendicular to these plates, we encounter plane stress approximation frequently.

2.4.1 Curved Panels

If the curvature of a thin panel is small, we may, locally, use a Cartesian system with coordinates x, s, and z, where the s coordinate changes in the direction of

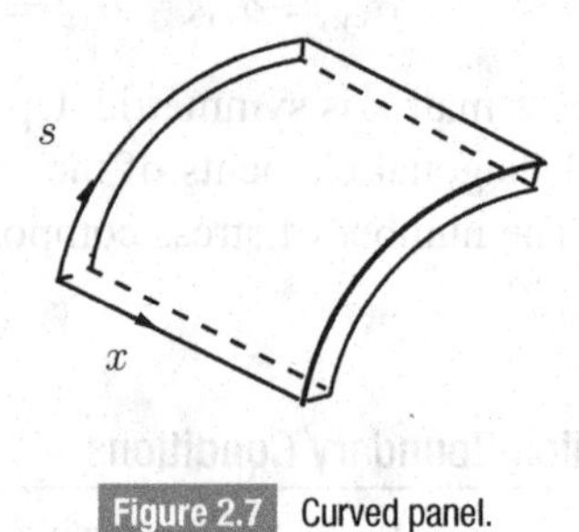

Figure 2.7 Curved panel.

curvature. This is shown in Fig. 2.7. We can adapt the plane stress equations with y replaced by s.

Then

$$\sigma_{xx,x} + \sigma_{xs,s} + f_x = 0, \tag{2.29}$$

$$\sigma_{xs,x} + \sigma_{ss,s} + f_s = 0, \tag{2.30}$$

and the boundary conditions

$$\sigma_{xx} n_x + \sigma_{xs} n_s = \bar{T}_x^n, \tag{2.31}$$

$$\sigma_{xs} n_x + \sigma_{ss} n_s = \bar{T}_s^n. \tag{2.32}$$

2.5 Transformation of Vectors

In the z-plane, if we introduce a new coordinate system x', y' with the same origin as that of the x, y system, but rotated by angle θ, counterclockwise, the components of a vector $\boldsymbol{v}$ will be different in the two systems. As shown in

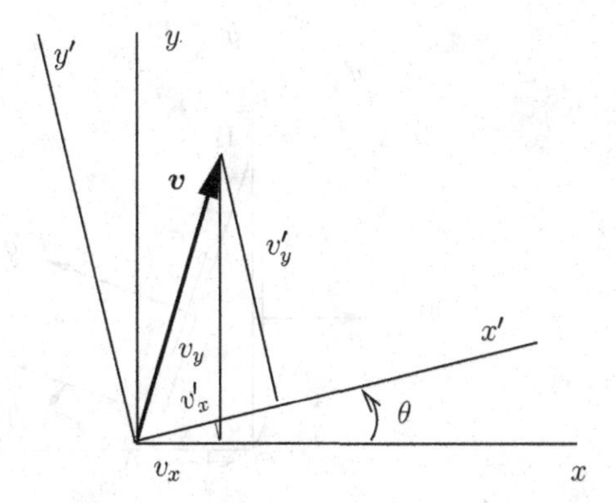

Figure 2.8 Vector components and coordinate rotation.

Fig. 2.8, let v_x and v_y be the components in the x, y-system and v'_x and v'_y be in the x', y' system. From geometry, we obtain the relations

$$v'_x = v_x \cos\theta + v_y \sin\theta,$$
$$v'_y = -v_x \sin\theta + v_y \cos\theta. \tag{2.33}$$

If we replace the vector $\boldsymbol{v}$ by the position vector

$$\boldsymbol{r} = x\boldsymbol{i} + y\boldsymbol{j}, \tag{2.34}$$

we find that its components in the two systems satisfy

$$x' = x \cos\theta + y \sin\theta,$$
$$y' = -x \sin\theta + y \cos\theta. \tag{2.35}$$

We mentioned earlier that the stress matrix is also called the stress tensor. In the plane stress case it is a 2 × 2-matrix,

$$[\boldsymbol{\sigma}] = \begin{bmatrix} \sigma_{xx} & \sigma_{xy} \\ \sigma_{xy} & \sigma_{yy} \end{bmatrix}. \tag{2.36}$$

In the rotated coordinate system, we expect the stress matrix to have the form

$$[\boldsymbol{\sigma}'] = \begin{bmatrix} \sigma'_{xx} & \sigma'_{xy} \\ \sigma'_{xy} & \sigma'_{yy} \end{bmatrix}. \tag{2.37}$$

2.6 Transformation of Tensors

In order to see how the stress components change as we rotate the coordinate system, we consider the static balance of a wedge of the stressed material, as shown in Fig. 2.9.

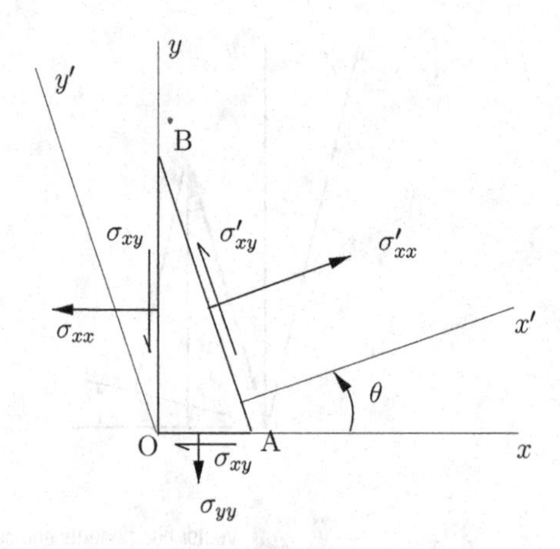

Figure 2.9 Stress transformation using static equilibrium.

We assume unit thickness for the wedge OAB in Fig. 2.9 and, first, balance the forces along the x' and y'-directions to get

$$\sigma'_{xx}AB = (\sigma_{xx}OB + \sigma_{xy}OA)\cos\theta + (\sigma_{xy}OB + \sigma_{yy}OA)\sin\theta, \quad (2.38)$$

$$\sigma'_{xy}AB = (\sigma_{yy}OA + \sigma_{xy}OB)\cos\theta - (\sigma_{xx}OB + \sigma_{xy}OS)\sin\theta. \quad (2.39)$$

Dividing by AB and noting

$$OB/AB = \cos\theta \equiv c, \quad OA/AB = \sin\theta \equiv s,$$

we have

$$\sigma'_{xx} = \sigma_{xx}c^2 + \sigma_{yy}s^2 + 2\sigma_{xy}cs,$$
$$\sigma'_{xy} = (\sigma_{yy} - \sigma_{xx})cs + \sigma_{xy}(c^2 - s^2).$$

The stress component σ'_{yy} is the normal stress on a plane 90° from the x'-plane. If we replace θ by $\theta + 90°$, we get

$$\cos(\theta + 90°) = -s, \quad \sin(\theta + 90°) = c, \quad (2.40)$$

and

$$\sigma'_{yy} = \sigma_{xx}s^2 + \sigma_{yy}c^2 - 2\sigma_{xy}cs.$$

Thus,

$$\begin{aligned}\sigma'_{xx} &= \sigma_{xx}c^2 + \sigma_{yy}s^2 + 2\sigma_{xy}cs,\\ \sigma'_{yy} &= \sigma_{xx}s^2 + \sigma_{yy}c^2 - 2\sigma_{xy}cs.\\ \sigma'_{xy} &= (\sigma_{yy} - \sigma_{xx})cs + \sigma_{xy}(c^2 - s^2).\end{aligned} \quad (2.41)$$

In comparison to the vector component transformation, Eq. (2.33), here we have a more complex relation among the components of stress. This relation applies to tensor components. To be precise, these relations apply to tensor components of rank two in two dimensions. A tensor of rank r has d^r components in d dimensions.

With the column vectors $\boldsymbol{v}$, $\boldsymbol{v}'$ and the rotation matrix $[\boldsymbol{Q}]$, defined as

$$\boldsymbol{v} = \begin{Bmatrix} v_x \\ v_y \end{Bmatrix}, \quad \boldsymbol{v}' = \begin{Bmatrix} v'_x \\ v'_y \end{Bmatrix}, \quad [\boldsymbol{Q}] = \begin{bmatrix} c & s \\ -s & c \end{bmatrix}, \tag{2.42}$$

we can verify the relations

$$\boldsymbol{v}' = [\boldsymbol{Q}]\boldsymbol{v}, \quad [\sigma'] = [\boldsymbol{Q}][\sigma][\boldsymbol{Q}]^T. \tag{2.43}$$

Using the double-angle relations

$$c^2 = \frac{1+\cos 2\theta}{2}, \quad s^2 = \frac{1-\cos 2\theta}{2}, \quad sc = \frac{\sin 2\theta}{2}, \tag{2.44}$$

the stress transformations relations can be written as

$$\begin{aligned} \sigma'_{xx} &= \sigma_m + \frac{\sigma_{xx}-\sigma_{yy}}{2}\cos 2\theta + \sigma_{xy}\sin 2\theta, \\ \sigma'_{yy} &= \sigma_m - \frac{\sigma_{xx}-\sigma_{yy}}{2}\cos 2\theta - \sigma_{xy}\sin 2\theta, \\ \sigma'_{xy} &= \frac{\sigma_{yy}-\sigma_{xx}}{2}\sin 2\theta + \sigma_{xy}\cos 2\theta, \end{aligned} \tag{2.45}$$

where the mean stress σ_m is defined as

$$\sigma_m = \frac{\sigma_{xx}+\sigma_{yy}}{2}. \tag{2.46}$$

Adding the first two equations in (2.45), we find

$$\sigma_m = \frac{\sigma'_{xx}+\sigma'_{yy}}{2}. \tag{2.47}$$

Thus, the mean stress remains an invariant under the coordinate rotation.

2.6.1 Principal Stresses

Having a formula for the normal stress on an arbitrary plane, we are interested in finding the angle θ that makes the normal stress a maximum (or a minimum). For this, we let

$$\frac{d\sigma'_{xx}}{d\theta} = 0.$$

This gives

$$\frac{\sigma_{xx}-\sigma_{yy}}{2}\sin 2\theta - \sigma_{xy}\cos 2\theta = 0.$$

Denoting the solution of this equation by θ_p, we get

$$\tan 2\theta_p = \frac{\sigma_{xy}}{(\sigma_{xx}-\sigma_{yy})/2}. \tag{2.48}$$

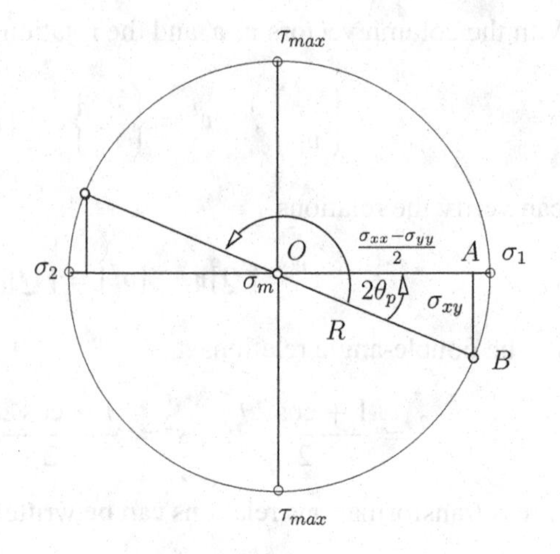

Figure 2.10 Angles of principal planes.

This equation has two solutions for $2\theta_p$, which are separated by 180°. Before we discuss these solutions fully, we observe from Eq. (2.45),

$$\frac{\sigma'_{xx}-\sigma'_{yy}}{2}=\frac{\sigma_{xx}-\sigma_{yy}}{2}\cos 2\theta+\sigma_{xy}\sin 2\theta,$$

$$\sigma'_{xy}=\frac{\sigma_{yy}-\sigma_{xx}}{2}\sin 2\theta+\sigma_{xy}\cos 2\theta.$$

Squaring and adding these two relations we get another invariant,

$$R^2=\left[\frac{\sigma'_{xx}-\sigma'_{yy}}{2}\right]^2+[\sigma'_{xy}]^2=\left[\frac{\sigma_{xx}-\sigma_{yy}}{2}\right]^2+[\sigma_{xy}]^2. \tag{2.49}$$

Here, we see another invariant of the tensor transformation. As shown in Fig. 2.10, the angle in Eq. (2.48) can be expressed two ways:

$$\cos 2\theta_p=\frac{\sigma_{xx}-\sigma_{yy}}{2R},\quad \sin 2\theta_p=\frac{\sigma_{xy}}{R}, \tag{2.50}$$

or

$$\cos 2\theta_p=-\frac{\sigma_{xx}-\sigma_{yy}}{2R},\quad \sin 2\theta_p=-\frac{\sigma_{xy}}{R}. \tag{2.51}$$

Substituting these in the equation for σ'_{xx} in Eq. (2.45), we get the maximum and minimum values,

$$\sigma_1=\sigma_m+R,\quad \sigma_2=\sigma_m-R. \tag{2.52}$$

These are the principal stresses; their planes, the principal planes, will be referred as p-planes.

Substituting the angles in the equation for the shear stress σ'_{xy} shows

$$\sigma'_{xy}=0. \tag{2.53}$$

Thus, the principal planes have pure normal stresses.

2.6.2 Maximum Shear Stress

We use the same procedure to maximize the shear stress σ'_{xy} in Eq. (2.45) to get

$$\sigma_{xy} \cos 2\theta + \frac{\sigma_{xx} - \sigma_{yy}}{2} \sin 2\theta = 0. \tag{2.54}$$

The solution of this equation is denoted by θ_s for the maximum shear planes (s-planes); and we have

$$\tan 2\theta_s = -\frac{\sigma_{xx} - \sigma_{yy}}{2\sigma_{xy}}. \tag{2.55}$$

This also has two solutions θ_s at 90° apart. We observe

$$\tan 2\theta_s \tan 2\theta_p = -1, \tag{2.56}$$

which is the condition for $2\theta_s$ to differ from $2\theta_p$ by 90° (or 270°).

When we substitute

$$\sin 2\theta_s = \frac{\sigma_{xx} - \sigma_{yy}}{2R}, \quad \cos 2\theta_p = -\frac{\sigma_{xy}}{R}, \tag{2.57}$$

in σ'_{xy} and σ'_{xx} we get

$$\sigma'_{xy} = R, \quad \sigma'_{xx} = \sigma_m. \tag{2.58}$$

2.6.3 Mohr's Circle

A visual representation of the state of plane stress at a point can be obtained using Mohr's circle. First, we note from the stress transformation relations, Eq. (2.45), that

$$(\sigma'_{xx} - \sigma_m)^2 + (\sigma'_{xy})^2 = R^2, \tag{2.59}$$

which shows that the normal stress $\sigma = \sigma'_{xx}$ and the shear stress $\tau = \sigma'_{xy}$ are coordinates of a point lying on a circle of radius R and center at $(\sigma_m, 0)$ in a σ, τ-plane. The principal stresses σ_1 and σ_2 are the extreme values of σ along a diameter where $\tau = 0$. The maximum shear stresses correspond to points $(\sigma_m, \pm R)$.

We maintain the convention for measuring angles as counterclockwise being positive. Then, as shown in Fig. 2.10, the state of stress on the x-plane must be below the point corresponding to $(\sigma_1, 0)$, subtending an angle $2\theta_p$ at the center. The sign convention for shear stresses is as follows: If we view the shear stress from below the plane to see it creating a clockwise moment then it is positive. On the x-plane σ_{xy} creates a counterclockwise moment and hence it is negative. We have to keep in mind that in the Cartesian system we see two planes x and y but on Mohr's circle each point on the circle is one plane. While the real planes differ by an angle θ, they differ by 2θ on Mohr's circle. As shown in

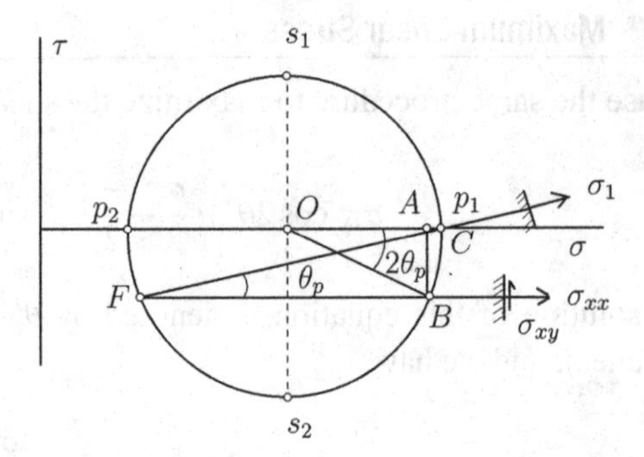

Figure 2.11 Mohr's circle showing principal stress planes, p_1, p_2, maximum shear stress planes, s_1, s_2, and the x-plane.

Fig. 2.11, the arc BC subtends $2\theta_p$ at the center and θ_p at F on the circumference. We can use the point F, called the focus, to get real angles.

2.6.4 Example: Principal Stresses and the Maximum Shear Stresses

Consider the state of stress shown in Fig. 2.12. Obtain the principal stresses and their planes and the maximum shear stresses and their planes. Also, include the

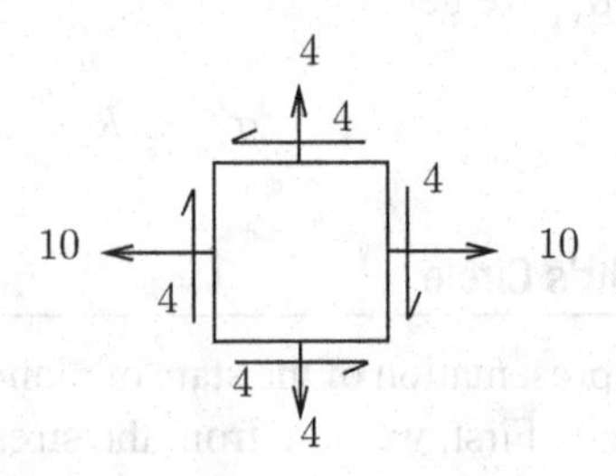

Figure 2.12 Stresses on an element.

x and y planes. All stresses are in MPas. The center of the circle is at

$$\sigma_m = (10 + 4)/2 = 7,$$

and its radius is

$$R^2 = [(10 - 4)/2]^2 + 4^2, \quad R = 5.$$

We draw Mohr's circle as shown in Fig. 2.13. On the x-plane the normal stress is $\sigma = 10$ MPa, and the shear stress of $\sigma_{xy} = 4$ MPa creates a clockwise moment and, so, it is positive. The x-plane is located on the circle at (10, 4). We draw a line parallel to the σ axis to locate the focus F. The angle 2θ is found from

$$\tan 2\theta = \frac{\sigma_{xy}}{(\sigma_{xx} - \sigma_{yy})/2} = 4/3;$$

$$2\theta = 53.13^\circ, \quad \theta = 26.57^\circ.$$

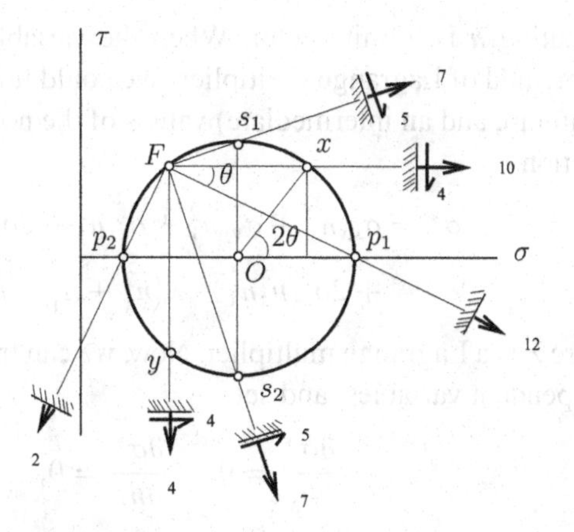

Figure 2.13 Mohr's circle for the example problem.

Connecting the focus F to s_1 and s_2 we observe the maximum shear planes and their angles. Similarly, connecting F to p_1 and p_2 we find the maximum and minimum principal stress planes. The steps taken in constructing the Mohr's circle can be listed as follows:

1. Find the radius R and the center $(\sigma_m, 0)$ of the circle to draw the circle in a (σ, τ) plane;
2. Locate the x-plane on the circle at the point $x : (\sigma_{xx}, \tau)$. This point is on the upper side if τ creates a clockwise moment;
3. Locate points corresponding to the maximum and the minimum values of σ, labeled p_1 and p_2, which are the principal planes;
4. Locate the maximum shear stress planes, s_1 and s_2;
5. From the point x draw a parallel to the σ-axis to intersect the circle at the focus, F, which would be a reflection of x about a vertical diameter;
6. Connect F to the y-plane point y, the principal planes p_1 and p_2, and the maximum shear planes s_1 and s_2;
7. Compute $2\theta_p$, the angle between x and p_1 subtended at the center;
8. Indicate the stresses on all of the planes.

2.7 Principal Stresses in Three Dimensions

From our discussion of the traction vector, we saw the normal stress on a plane with normal $\boldsymbol{n}$ is given by Eq. (2.13) as

$$\begin{aligned}\sigma &= \{\boldsymbol{n}\}^T[\sigma]\{\boldsymbol{n}\} \\ &= \sigma_{xx}n_x^2 + \sigma_{yy}n_y^2 + \sigma_{zz}n_z^2 + 2\sigma_{xy}n_xn_y + 2\sigma_{yz}n_yn_z + 2\sigma_{zx}n_zn_x. \qquad (2.60)\end{aligned}$$

This quantity can be considered as a function of the variables: n_x, n_y, and n_z. However, these three variables are not independent due to the constraint

$$n_x^2 + n_y^2 + n_z^2 = 1, \qquad (2.61)$$

indicating $\boldsymbol{n}$ is a unit vector. When the variables are not independent, we use the method of Lagrange multiplier. We could find the extremum (i.e., maximum, minimum, and an intermediate) values of the normal stress σ using the modified function

$$\sigma^* = \sigma_{xx}n_x^2 + \sigma_{yy}n_y^2 + \sigma_{zz}n_z^2 + 2\sigma_{xy}n_xn_y + 2\sigma_{yz}n_yn_z + 2\sigma_{zx}n_zn_x - \lambda\left(n_x^2 + n_y^2 + n_z^2 - 1\right), \tag{2.62}$$

where λ is a Lagrange multiplier. Now, we can treat the components of $\boldsymbol{n}$ as three independent variables, and set

$$\frac{\partial \sigma^*}{\partial n_x} = 0, \quad \frac{\partial \sigma^*}{\partial n_y} = 0, \quad \frac{\partial \sigma^*}{\partial n_z} = 0. \tag{2.63}$$

These equations are

$$\begin{aligned} \sigma_{xx}n_x + \sigma_{xy}n_y + \sigma_{xz}n_z &= \lambda n_x, \\ \sigma_{xy}n_x + \sigma_{yy}n_y + \sigma_{yz}n_z &= \lambda n_y, \\ \sigma_{xz}n_x + \sigma_{yz}n_y + \sigma_{zz}n_z &= \lambda n_z. \end{aligned} \tag{2.64}$$

We may express these in the compact form

$$\begin{bmatrix} \sigma_{xx} - \lambda & \sigma_{xy} & \sigma_{xz} \\ \sigma_{xy} & \sigma_{yy} - \lambda & \sigma_{yz} \\ \sigma_{xz} & \sigma_{yz} & \sigma_{zz} - \lambda \end{bmatrix} \begin{Bmatrix} n_x \\ n_y \\ n_z \end{Bmatrix} = \begin{Bmatrix} 0 \\ 0 \\ 0 \end{Bmatrix}. \tag{2.65}$$

This system of homogeneous equations has a nonzero solution $\boldsymbol{n}$ if the determinant of the system is zero (i.e., the three equations are not independent). This gives

$$\begin{vmatrix} \sigma_{xx} - \lambda & \sigma_{xy} & \sigma_{xz} \\ \sigma_{xy} & \sigma_{yy} - \lambda & \sigma_{yz} \\ \sigma_{xz} & \sigma_{yz} & \sigma_{zz} - \lambda \end{vmatrix} = 0. \tag{2.66}$$

This is what is called an eigenvalue problem. By expanding the determinant, we get a cubic equation for λ,

$$-\lambda^3 + I_1\lambda^2 - I_2\lambda + I_3 = 0, \tag{2.67}$$

where I_1, I_2, and I_3 are called the invariants of the stress matrix. The roots of the cubic depend only on the invariants. If we multiply the first equation in Eq. (2.64) by n_x and the second by n_y and the third by n_z and add all the three, we find that the extreme values of σ in Eq. (2.60) satisfy

$$\sigma = \lambda. \tag{2.68}$$

Thus, the three roots of the cubic are indeed the principal stresses. The principal stresses are independent of the coordinate system used, and they depend on I_1, I_2, and I_3, only. From this we can appreciate why these coefficients are called the invariants of the matrix.

Let σ_1 be a principal stress. Substituting this for λ in Eq. (2.65), we get three homogeneous equations for n_x, n_y, and n_z. The fact that the determinant of this system is zero implies that at least one of these equations is a linear combination of the other two. Then, assuming a value for one of the components of $\boldsymbol{n}$, we can solve for the other two. We can make the vector obtained a unit vector by dividing by its length. This operation is called normalization. This unit vector represents the principal plane corresponding to the principal stress σ_1. We call this an eigenvector corresponding to the eigenvalue σ_1. This process can be repeated for the other two eigenvalues. The theory of matrices guarantees three orthogonal eigenvectors for a symmetric 3×3 matrix.

2.7.1 Example: Eigenvalues and Eigenvectors

Find the principal stresses and their directions for the stress matrix

$$[\sigma] = \begin{bmatrix} 8 & 0 & 4 \\ 0 & 4 & 0 \\ 4 & 0 & 2 \end{bmatrix}.$$

Replacing λ by σ, the principal stress, we have to set

$$\begin{vmatrix} 8-\sigma & 0 & 4 \\ 0 & 4-\sigma & 0 \\ 4 & 0 & 2-\sigma \end{vmatrix} = 0.$$

Expanding this,

$$(8-\sigma)[(4-\sigma)(2-\sigma)] - 0 + 4[-(4-\sigma)4] = 0.$$

Removing the common factor $(4-\sigma)$, we get

$$(8-\sigma)(2-\sigma) - 4^2 = 0, \quad \sigma^2 - 10\sigma = 0.$$

Thus the roots are

$$\sigma_1 = 10, \quad \sigma_2 = 4, \quad \sigma_3 = 0.$$

These are the principal stresses. To find their directions, we begin by substituting $\lambda = \sigma = 10$ in Eq. (2.65). This gives

$$\begin{aligned} -2n_x + 4n_z &= 0, \\ -6n_y &= 0, \\ 4n_x - 8n_z &= 0. \end{aligned}$$

The vector (2, 0, 1) would satisfy these equations. After normalizing, we get

$$\boldsymbol{n}^{(1)} = \frac{1}{\sqrt{5}} \begin{Bmatrix} 2 \\ 0 \\ 1 \end{Bmatrix}.$$

Similarly, for $\sigma_2 = 4$, we get

$$\boldsymbol{n}^{(2)} = \begin{Bmatrix} 0 \\ 1 \\ 0 \end{Bmatrix},$$

and for $\sigma_3 = 0$,

$$\boldsymbol{n}^{(3)} = \frac{1}{\sqrt{5}} \begin{Bmatrix} 1 \\ 0 \\ -2 \end{Bmatrix}.$$

It is easy to verify that these three normals are orthogonal to each other.

This example is a partially hidden plane stress problem (more exactly, a generalized plane stress problem). There are a number of software functions that will give the eigenvalues and eigenfunctions of given matrices.

Once the three unit vectors, $\boldsymbol{n}^{(1)}$, $\boldsymbol{n}^{(2)}$, and $\boldsymbol{n}^{(3)}$, are known, we may choose a coordinate system oriented along these vectors. In this system the stress matrix has no shear stress components and we get the diagonal matrix

$$[\sigma] = \begin{bmatrix} \sigma_1 & 0 & 0 \\ 0 & \sigma_2 & 0 \\ 0 & 0 & \sigma_3 \end{bmatrix}. \tag{2.69}$$

We may construct three Mohr's circles – one on each plane. If we superpose these three, it would look like the sketch in Fig. 2.14.

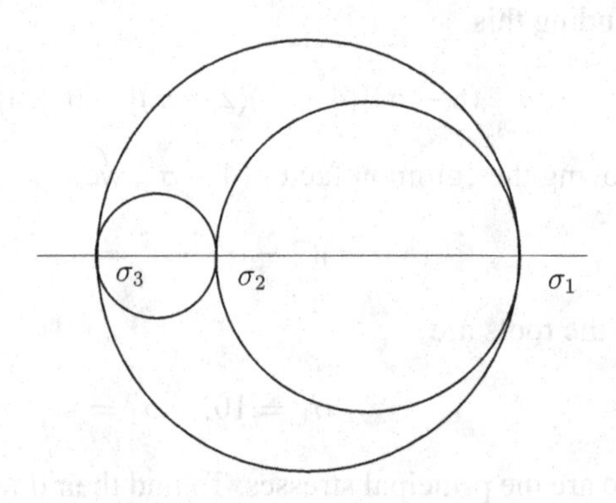

Figure 2.14 Superposed Mohr's circles for the three-dimensional case with $\sigma_1 > \sigma_2 > \sigma_3$.

Each Mohr's circle has its own maximum shear stress, which is its radius. In general, the greatest value for shear can be expressed as

$$\tau_{max} = \frac{|\sigma_1 - \sigma_2|}{2} \quad \text{or} \quad \frac{|\sigma_2 - \sigma_3|}{2} \quad \text{or} \quad \frac{|\sigma_3 - \sigma_1|}{2}. \tag{2.70}$$

2.8 Displacement and Strain

With respect to a fixed Cartesian system, x, y, z, a body occupying a volume B in space moves to the position B' under applied loads. A material particle P in B moves to P' (see Fig. 2.15). The vector $\overrightarrow{PP'}$ is called the displacement vector, $\boldsymbol{u}(x, y, z)$, which has the component form

$$\boldsymbol{u} = u\boldsymbol{i} + v\boldsymbol{j} + w\boldsymbol{k}. \tag{2.71}$$

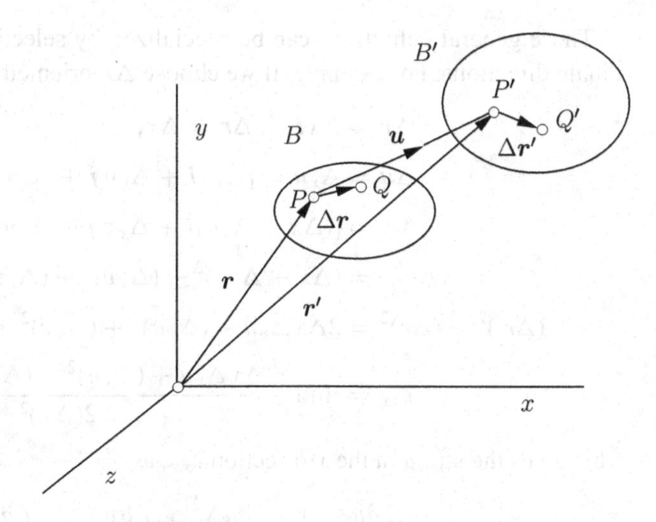

Figure 2.15 Displacement vector for deformation.

The position vector $\boldsymbol{r}$ is given by

$$\boldsymbol{r} = x\boldsymbol{i} + y\boldsymbol{j} + z\boldsymbol{k}. \tag{2.72}$$

The small element $\overrightarrow{PQ} = \Delta\boldsymbol{r}$ becomes $\overrightarrow{P'Q'} = \Delta\boldsymbol{r}'$ during the deformation process. Noting $\overrightarrow{QQ'} = \boldsymbol{u} + \Delta\boldsymbol{u}$, we can express the deformed length

$$\Delta\boldsymbol{r}' = \Delta\boldsymbol{r} + \boldsymbol{u} + \Delta\boldsymbol{u} - \boldsymbol{u} = \Delta\boldsymbol{r} + \Delta\boldsymbol{u}. \tag{2.73}$$

In order to account for large displacement gradients and large rotation of elements we have to modify the definition of normal strain from the simple form of "change in length over original length." The normal strain of an element $\Delta\boldsymbol{r}$ oriented in the direction of the unit vector $\boldsymbol{n}$ is defined by

$$\epsilon_{nn} = \lim_{\Delta r \to 0} \frac{(\Delta r')^2 - (\Delta r)^2}{2(\Delta r)^2}, \tag{2.74}$$

where Δr is the magnitude of the vector $\Delta\boldsymbol{r}$, and so on. When the deformed length is close to the original length and if the rotation of the element is small, this definition reduces to the classical definition. To calculate the shear strain as a change in angle between two elements that are initially at 90°, we choose two orthogonal elements $\boldsymbol{r}_1$ and $\boldsymbol{r}_2$ along the unit vectors $\boldsymbol{n}$ and $\boldsymbol{m}$ and define

$$\epsilon_{nm} = \epsilon_{mn} = \lim_{\Delta r_1 \to 0, \Delta r_2 \to 0} \frac{\Delta\boldsymbol{r}_1' \bullet \Delta\boldsymbol{r}_2'}{2\Delta r_1 \Delta r_2}. \tag{2.75}$$

The dot product in this expression will be zero if the deformed elements are still at 90° and we get zero shear strain.

These general definitions can be specialized by selecting $\Delta \boldsymbol{r}$ along the coordinate directions. For example, if we choose $\Delta \boldsymbol{r}$ oriented along the x-direction,

$$\Delta \boldsymbol{r} = \Delta x \boldsymbol{i}, \quad \Delta r = \Delta x,$$

$$\Delta \boldsymbol{u} = \Delta_x \boldsymbol{u} = [\Delta_x u \boldsymbol{i} + \Delta_x v \boldsymbol{j} + \Delta_x w \boldsymbol{k}],$$

$$\Delta \boldsymbol{r}' = [(\Delta x + \Delta_x u)\boldsymbol{i} + \Delta_x v \boldsymbol{j} + \Delta_x w \boldsymbol{k}],$$

$$(\Delta r')^2 = (\Delta x + \Delta_x u)^2 + (\Delta_x v)^2 + (\Delta_x w)^2,$$

$$(\Delta r')^2 - (\Delta r)^2 = 2\Delta x \Delta_x u + (\Delta_x u)^2 + (\Delta_x v)^2 + (\Delta_x w)^2,$$

$$\epsilon_{xx} = \lim_{\Delta x \to 0} \frac{2\Delta x \Delta_x u + (\Delta_x u)^2 + (\Delta_x v)^2 + (\Delta_x w)^2}{2(\Delta x)^2}.$$

This gives the strain in the x direction ϵ_{xx}, as

$$\epsilon_{xx} = \frac{\partial u}{\partial x} + \frac{1}{2}\left[\left(\frac{\partial u}{\partial x}\right)^2 + \left(\frac{\partial v}{\partial x}\right)^2 + \left(\frac{\partial w}{\partial x}\right)^2\right]. \tag{2.76}$$

Similarly,

$$\epsilon_{yy} = \frac{\partial v}{\partial y} + \frac{1}{2}\left[\left(\frac{\partial u}{\partial y}\right)^2 + \left(\frac{\partial v}{\partial y}\right)^2 + \left(\frac{\partial w}{\partial y}\right)^2\right].$$

$$\epsilon_{zz} = \frac{\partial w}{\partial z} + \frac{1}{2}\left[\left(\frac{\partial u}{\partial z}\right)^2 + \left(\frac{\partial v}{\partial z}\right)^2 + \left(\frac{\partial w}{\partial z}\right)^2\right].$$

To obtain the shear strain between the x and y directions, we choose

$$\Delta \boldsymbol{r}_1 = \Delta x \boldsymbol{i}, \quad \Delta \boldsymbol{r}_2 = \Delta y \boldsymbol{j}. \tag{2.77}$$

These give

$$\Delta \boldsymbol{r}_1' = \Delta x \boldsymbol{i} + \Delta_x \boldsymbol{u}, \quad \Delta \boldsymbol{r}_2' = \Delta y \boldsymbol{j} + \Delta_y \boldsymbol{u}. \tag{2.78}$$

Using this in the shear strain formula, Eq. (2.75), we find

$$\epsilon_{xy} = \frac{1}{2}\left[\frac{\partial u}{\partial y} + \frac{\partial v}{\partial x} + \frac{\partial u}{\partial x}\frac{\partial u}{\partial y} + \frac{\partial v}{\partial x}\frac{\partial v}{\partial y} + \frac{\partial w}{\partial x}\frac{\partial w}{\partial y}\right] \tag{2.79}$$

Similar calculations using element vectors along the y and z directions and z and x directions result in

$$\epsilon_{yz} = \frac{1}{2}\left[\frac{\partial v}{\partial z} + \frac{\partial w}{\partial y} + \frac{\partial u}{\partial y}\frac{\partial u}{\partial z} + \frac{\partial v}{\partial y}\frac{\partial v}{\partial z} + \frac{\partial w}{\partial y}\frac{\partial w}{\partial z}\right], \tag{2.80}$$

$$\epsilon_{zx} = \frac{1}{2}\left[\frac{\partial w}{\partial x} + \frac{\partial u}{\partial z} + \frac{\partial u}{\partial y}\frac{\partial u}{\partial z} + \frac{\partial v}{\partial y}\frac{\partial v}{\partial z} + \frac{\partial w}{\partial y}\frac{\partial w}{\partial z}\right]. \tag{2.81}$$

We note that the shear strains are symmetric by definition, that is, $\epsilon_{xy} = \epsilon_{yx}$, and so on. These nonlinear, large strains are known as the Green-Lagrange strains. We may form a strain matrix

$$[\boldsymbol{\epsilon}] = \begin{bmatrix} \epsilon_{xx} & \epsilon_{xy} & \epsilon_{xz} \\ \epsilon_{xy} & \epsilon_{yy} & \epsilon_{yz} \\ \epsilon_{xz} & \epsilon_{yz} & \epsilon_{zz} \end{bmatrix}. \tag{2.82}$$

When the magnitude of the displacement derivatives and the rotation of the elements during deformation are small, we neglect the nonlinear terms and use the linear small strain expressions

$$\epsilon_{xx} = \frac{\partial u}{\partial x} \quad \epsilon_{yy} = \frac{\partial v}{\partial y} \quad \epsilon_{zz} = \frac{\partial w}{\partial z}, \tag{2.83}$$

$$\epsilon_{xy} = \frac{1}{2}\left[\frac{\partial u}{\partial y} + \frac{\partial v}{\partial x}\right], \epsilon_{yz} = \frac{1}{2}\left[\frac{\partial v}{\partial z} + \frac{\partial w}{\partial y}\right] \quad \epsilon_{zx} = \frac{1}{2}\left[\frac{\partial w}{\partial x} + \frac{\partial u}{\partial z}\right]. \tag{2.84}$$

If we superimpose the origins of the undeformed elements $\Delta x \boldsymbol{i}$ and $\Delta y \boldsymbol{j}$ and their deformed images $\Delta x \boldsymbol{i} + \Delta_x \boldsymbol{u}$ and $\Delta y \boldsymbol{j} + \Delta_y \boldsymbol{u}$ we get the picture shown in Fig. 2.16. From this, we can see the geometrical meanings of the small normal strains as change in length over original length and shear strains as half of the reduction in the included angle from the original $\pi/2$ radians.

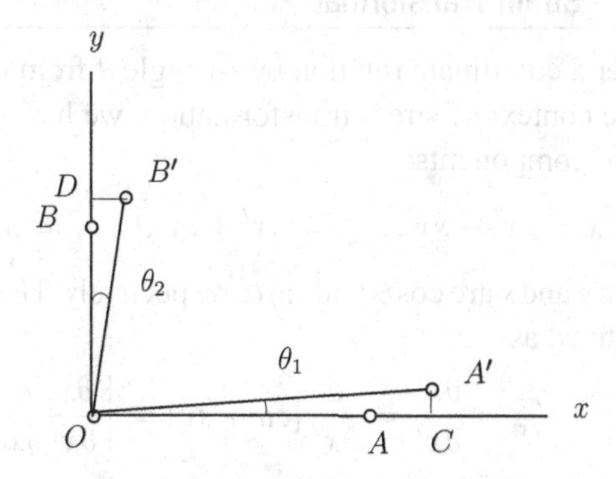

Figure 2.16 Superposition of elements OA and OB before deformation with O′A′ and O′B′ after deformation.

We have

$$OA = \Delta x, \quad OA' \approx OC = \Delta x\left(1 + \frac{\partial u}{\partial x}\right),$$

$$OB = \Delta y, \quad OB' \approx OD = \Delta y\left(1 + \frac{\partial v}{\partial y}\right),$$

$$\epsilon_{xx} = \frac{OC - OA}{OA} = \frac{\partial u}{\partial x} \quad \epsilon_{yy} = \frac{OD - OB}{OB} = \frac{\partial v}{\partial y}.$$

$$\theta_1 = \frac{CA'}{OC} = \frac{\partial v}{\partial x} \quad \theta_2 = \frac{DB'}{OD} = \frac{\partial u}{\partial y}.$$

$$\epsilon_{xy} = \frac{\theta_1 + \theta_2}{2} = \frac{1}{2}\left[\frac{\partial u}{\partial y} + \frac{\partial v}{\partial x}\right].$$

The shear strains we have defined are known as *mathematical shear strains*, which are calculated using the average of θ_1 and θ_2. The *engineering shear strains* are defined as the total change is angle, $\theta_1 + \theta_2$. These are denoted by γ_{xy}, and so on. We have

$$\gamma_{xy} = 2\epsilon_{xy}, \quad \gamma_{yz} = 2\epsilon_{yz}, \quad \gamma_{zx} = 2\epsilon_{zx}. \tag{2.85}$$

2.8.1 Plane Strain

If

$$\epsilon_{zz} = 0, \quad \epsilon_{zx} = 0, \quad \epsilon_{zy} = 0,$$

the state of strain is referred to as *plane strain*. We may visualize a body kept at constant length in the z-direction using rigid blocks and with a uniform cross section. The loading on the surface should not vary along z to achieve plane strain conditions. In this case, elasticity problems involve three nonzero strains: ϵ_{xx}, ϵ_{yy}, and ϵ_{xy}, given by

$$\epsilon_{xx} = \frac{\partial u}{\partial x}, \quad \epsilon_{yy} = \frac{\partial v}{\partial y}, \quad \epsilon_{xy} = \frac{1}{2}\left(\frac{\partial u}{\partial y} + \frac{\partial v}{\partial x}\right). \tag{2.86}$$

2.8.2 Strain Transformation

Under a coordinate rotation by an angle θ from the x, y-system to x', y'-system, in the context of stress transformation, we have seen the transformation law for vector components:

$$x = cx' - sy', \quad y = sx' + cy', \quad u' = cu + sv, \quad v' = -su + cv,$$

where c and s are $\cos\theta$ and $\sin\theta$, respectively. The normal strain in the x'-direction is defined as

$$\epsilon'_{xx} = \frac{\partial u'}{\partial x'} = \frac{\partial}{\partial x'}[cu + sv] = \left[\frac{\partial x}{\partial x'}\frac{\partial}{\partial x} + \frac{\partial y}{\partial x'}\frac{\partial}{\partial y}\right][cu + sv], \tag{2.87}$$

$$= \left[c\frac{\partial}{\partial x} + s\frac{\partial}{\partial y}\right][cu + sv], \tag{2.88}$$

$$= c^2\epsilon_{xx} + s^2\epsilon_{yy} + 2cs\epsilon_{xy}, \tag{2.89}$$

which, interestingly, is the same law for the stress components. This derivation is based on geometry whereas the stress transformation is based on statics. We may complete the list as

$$\epsilon'_{yy} = s^2\epsilon_{xx} + c^2\epsilon_{yy} - 2cs\epsilon_{xy}, \tag{2.90}$$

$$\epsilon'_{xy} = cs(\epsilon_{yy} - \epsilon_{xx}) + (c^2 - s^2)\epsilon_{xy}. \tag{2.91}$$

Because of this transformation law, we refer to the collection of strain components as the strain tensor.

In the laboratory, we measure surface strains using *strain rosettes*. These are three resistors attached to the surface in different orientations. When there is a normal strain along the length of a resistor, the length of the resistor is altered and this causes the resistance to change. By measuring the resistance, we can obtain the normal strain. Three resistors are needed to resolve ϵ_{xx}, ϵ_{xx}, and ϵ_{xy}.

Now, we can apply the same approach we used for stresses to find the directions and magnitudes of the maximum and minimum normal strains, which are called the principal strains, those of the maximum shear strain, and to draw Mohr's circle. The sign convention for shear strains in the Mohr's circle representation would be clear if we associate it with the shear stress. In the Cartesian form, we associate positive shear strain with positive shear stress, which for the x-plane is negative and for the y-plane is positive.

From the expression for ϵ'_{xx}, $\cos\theta$ and $\sin\theta$ are expressed as n_x and n_y, the components of a unit vector, the strain in the direction $\boldsymbol{n}$ can be written as

$$\epsilon_{nn} = \{n\}^T[\epsilon]\{n\}. \tag{2.92}$$

This is analogous to the normal stress expression obtained using the stress matrix and the unit vector.

2.8.3 Displacement Boundary Conditions

In elasticity problems, certain surfaces of the body may have prescribed displacements. For example, for a cantilever beam one end has zero displacements. In general, on certain portions of the boundary, we may have

$$u = \bar{u}, \quad v = \bar{v}, \tag{2.93}$$

where, as before, $\bar{u}$ and $\bar{v}$ are given functions.

2.8.4 Compatibility Equations

The strain-displacement relations, Eq. (2.86), can be viewed as three partial differential equations for two functions, u and v, if we know the strains. This leads to a redundancy in the number of equations. The solution to this puzzle is to recognize that the three strains are not independent. They satisfy a differential equation that is known as the compatibility equation. We have

$$\frac{\partial^2 \epsilon_{xx}}{\partial y^2} = \frac{\partial^3 u}{\partial x \partial y^2}, \quad \frac{\partial^2 \epsilon_{yy}}{\partial x^2} = \frac{\partial^3 v}{\partial x^2 \partial y}. \tag{2.94}$$

Adding these two relations, we obtain the compatibility equation,

$$\frac{\partial^2 \epsilon_{xx}}{\partial y^2} + \frac{\partial^2 \epsilon_{yy}}{\partial x^2} = 2\frac{\partial^2 \epsilon_{xy}}{\partial x \partial y}. \tag{2.95}$$

Thus, we cannot choose three strains arbitrarily; they have to satisfy the compatibility equation. The compatibility equation plays an important role when we start with assumed stresses and obtain the strains using Hooke's law. If compatibility equation is not satisfied the assumed stresses have to be revised to satisfy the compatibility and to get two displacements from integrating the three strains. In the three-dimensional case there are six compatibility equations.

2.9 Generalized Hooke's Law

The stress and strain descriptions we have seen so far are independent of the material. The relations between stress and strain are called constitutive relations. An elastic material is called linear elastic if the stresses are linearly related to the strains. Generalizing the simple one-dimensional case where stress is proportional to strain, we have, for a linear, isotropic, elastic material,

$$\epsilon_{xx} = \frac{1}{E}[\sigma_{xx} - \nu(\sigma_{yy} + \sigma_{zz})], \tag{2.96}$$

$$\epsilon_{yy} = \frac{1}{E}[\sigma_{yy} - \nu(\sigma_{zz} + \sigma_{xx})], \tag{2.97}$$

$$\epsilon_{zz} = \frac{1}{E}[\sigma_{zz} - \nu(\sigma_{xx} + \sigma_{yy})], \tag{2.98}$$

$$\gamma_{xy} = 2\epsilon_{xy} = \frac{\sigma_{xy}}{G}, \tag{2.99}$$

$$\gamma_{yz} = 2\epsilon_{yz} = \frac{\sigma_{yz}}{G}, \tag{2.100}$$

$$\gamma_{xz} = 2\epsilon_{xz} = \frac{\sigma_{xz}}{G}, \tag{2.101}$$

where E is the Young's modulus, ν is the Poisson's ratio, and G is the shear modulus, which can be written as

$$G = \frac{E}{2(1+\nu)}. \tag{2.102}$$

This set of relations with just two independent material constants is dictated by the requirement of isotropy. If we rotate the coordinate system, stresses and strains would change, but the Hooke's law should be independent of the coordinates used. We will discuss this further when we present anisotropic relations.

2.9.1 Hooke's Law for Plane Stress

Under the plane stress assumption $\sigma_{zz} = \sigma_{xz} = \sigma_{yz} = 0$, and Hooke's law reduces to

$$\epsilon_{xx} = \frac{1}{E}[\sigma_{xx} - \nu\sigma_{yy}], \tag{2.103}$$

$$\epsilon_{yy} = \frac{1}{E}[\sigma_{yy} - \nu\sigma_{xx})], \tag{2.104}$$

$$\gamma_{xy} = \frac{1}{G}\sigma_{xy}, \tag{2.105}$$

The strain in the z-direction can be obtained as

$$\epsilon_{zz} = -\frac{\nu}{E}[\sigma_{xx} + \sigma_{yy}]. \tag{2.106}$$

2.9.2 Hooke's Law for Plane Strain

From $\epsilon_{zz} = 0$, we find

$$\sigma_{zz} = \nu[\sigma_{xx} + \sigma_{yy}]. \tag{2.107}$$

We use this equation to eliminate σ_{zz} from the generalized Hooke's law to get

$$\epsilon_{xx} = \frac{1-\nu^2}{E}\left[\sigma_{xx} - \frac{\nu}{1-\nu}\sigma_{yy}\right], \tag{2.108}$$

$$\epsilon_{yy} = \frac{1-\nu^2}{E}\left[\sigma_{yy} - \frac{\nu}{1-\nu}\sigma_{xx})\right], \tag{2.109}$$

$$\gamma_{xy} = \frac{1}{G}\sigma_{xy}. \tag{2.110}$$

The plane stress and plane strain constitutive relations can be cast in a similar form if we introduce the effective Young's modulus E' and Poisson's ratio ν',

$$\begin{aligned} E' &= E/(1-\nu^2), & \nu' &= \nu/(1-\nu), & &\text{Plane strain,} \\ E' &= E, & \nu' &= \nu, & &\text{Plane stress.} \end{aligned} \tag{2.111}$$

This way, if we have a solution for a plane strain problem for a domain in the x, y-plane, we can transform it into a plane stress solution by changing the elastic constants.

2.9.3 Example: Cantilever Beam under Plane Stress

Let us consider a cantilever beam with length ℓ, height $2c$, and depth b subjected to a tip load P. If $b \ll c$, we may assume plane stress conditions with the z-axis

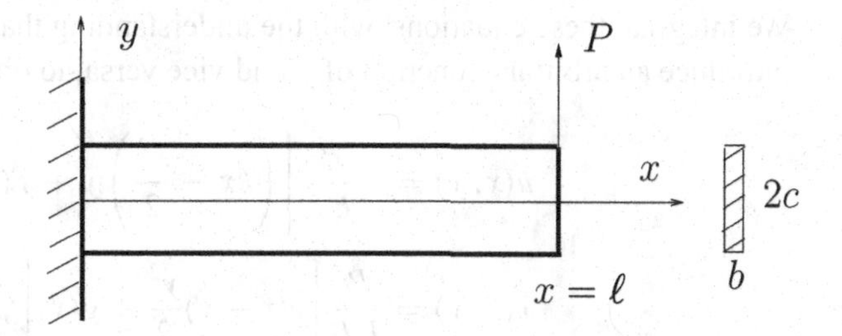

Figure 2.17 Cantilever beam under a tip load.

in the direction of the depth. We neglect gravitational forces. We begin with the elementary beam theory approximation

$$\sigma_{xx} = -\frac{My}{I}, \quad M = P(\ell - x), \quad I = \frac{2}{3}bc^3. \tag{2.112}$$

The equilibrium equation

$$\frac{\partial \sigma_{xx}}{\partial x} + \frac{\partial \sigma_{xy}}{\partial y} = 0, \tag{2.113}$$

gives

$$\frac{\partial \sigma_{xy}}{\partial y} = -\frac{P}{I}y, \tag{2.114}$$

which can be integrated to get

$$\sigma_{xy} = -\frac{P}{2I}y^2 + K, \tag{2.115}$$

where K is a constant. At the top and bottom surfaces of the beam there are no shear stresses. That is $\sigma_{xy} = 0$ when $y = \pm c$. Then $K = Pc^2/(2I)$ and

$$\sigma_{xy} = \frac{P}{2I}(c^2 - y^2). \tag{2.116}$$

This is the parabolic distribution of shear stress in the y direction. The second equilibrium equation in conjunction with the conditions $\sigma_{yy} = 0$ at $y = \pm c$ gives $\sigma_{yy} = 0$. Also, the tip load P is the resultant of this parabolically distributed shear stress. It can be seen

$$\frac{1}{2}\int_{-c}^{c}(c^2 - y^2)bdy = I. \tag{2.117}$$

From Hooke's law for plane stress,

$$\begin{aligned} \frac{\partial u}{\partial x} &= \frac{\sigma_{xx}}{E}, & \frac{\partial v}{\partial y} &= -\nu\frac{\sigma_{xx}}{E}, \\ \frac{\partial u}{\partial x} &= -\frac{P}{EI}(\ell - x)y, & \frac{\partial v}{\partial y} &= \frac{P}{EI}\nu(\ell - x)y. \end{aligned} \tag{2.118}$$

We integrate these equations, with the understanding that integration with x will introduce an arbitrary function of y and vice versa, to obtain

$$\begin{aligned} u(x, y) &= -\frac{P}{EI}\left[\left(\ell x - \frac{x^2}{2}\right)y + f(y)\right], \\ v(x, y) &= \frac{P}{EI}\left[\nu(\ell - x)\frac{y^2}{2} + g(x)\right], \end{aligned} \tag{2.119}$$

where $f(y)$ and $g(x)$ are the arbitrary functions. Next, we substitute u and v in the shear strain constitutive relation,

$$\gamma_{xy} = \frac{\sigma_{xy}}{G},$$

$$\frac{df}{dy} + \ell x - \frac{x^2}{2} - \frac{dg}{dx} + \nu\frac{y^2}{2} = -\frac{E}{G}\frac{c^2 - y^2}{2}. \tag{2.120}$$

Separating the variables,

$$\frac{df}{dy} = -\nu\frac{y^2}{2} - \frac{E}{G}\frac{c^2 - y^2}{2} + C, \tag{2.121}$$

$$\frac{dg}{dx} = \ell x - \frac{x^2}{2} + C, \tag{2.122}$$

where C is a new constant. We integrate these two equations

$$f = -\nu\frac{y^3}{6} - \frac{E}{G}\left[\frac{c^2 y}{2} - \frac{y^3}{6}\right] + Cy + D_1,$$

$$g = \frac{\ell x^2}{2} - \frac{x^3}{6} + Cx + D_2,$$

with two new constants of integration, D_1 and D_2. Substituting in the displacements, we find

$$u = -\frac{P}{EI}\left[\left(\ell x - \frac{x^2}{2}\right)y - \nu\frac{y^3}{6} - \frac{E}{G}\left(\frac{c^2 y}{2} - \frac{y^3}{6}\right) + Cy + D_1\right], \quad (2.123)$$

$$v = \frac{P}{EI}\left[\nu(\ell - x)\frac{y^2}{2} + \frac{\ell x^2}{2} - \frac{x^3}{6} + Cx + D_2\right]. \quad (2.124)$$

Although we would like to satisfy the boundary conditions at the fixed end, namely, $u = 0$, $v = 0$, at $x = 0$ for all y between $-c$ and c, these solutions do not allow it for all values of y. The next step is to try to meet these requirements at $x = 0$ and $y = 0$. This gives $D_1 = 0$ and $D_2 = 0$. That leaves C. The contributions due to C can be viewed as those due to a small counterclockwise rigid body rotation of the beam about the point $(0, 0)$. We may set

$$\frac{\partial u}{\partial y} = 0, \quad \text{at} \quad x = 0, \quad y = 0. \quad (2.125)$$

This leads to

$$C = \frac{Ec^2}{2G},$$

and to the final expressions

$$u = -\frac{P}{EI}\left[\left(\ell x - \frac{x^2}{2}\right)y - \nu\frac{y^3}{6} + \frac{Ey^3}{6G}\right], \quad (2.126)$$

$$v = \frac{P}{EI}\left[\nu(\ell - x)\frac{y^2}{2} + \frac{\ell x^2}{2} - \frac{x^3}{6} + \frac{Ec^2}{2G}x\right]. \quad (2.127)$$

The tip deflection of the beam measured at the point $x = \ell$, $y = 0$ is obtained as

$$v(\ell, 0) = \frac{P\ell^3}{3EI}\left[1 + \frac{3Ec^2}{2G\ell^2}\right]. \quad (2.128)$$

When we compare this with the elementary beam result, $P\ell^3/(3EI)$, we see an additional contribution, $3Ec^2/(2G\ell^2)$, which is attributed to the shear strains and called the shear correction. As it is proportional to $(c/\ell)^2$, we may neglect it when $\ell \gg c$. Recall, we are not able to satisfy the boundary conditions at the fixed end exactly. Generally, the shear correction factor is included as

$$v(\ell, 0) = \frac{P\ell^3}{3EI}\left[1 + k\frac{Ec^2}{G\ell^2}\right],$$

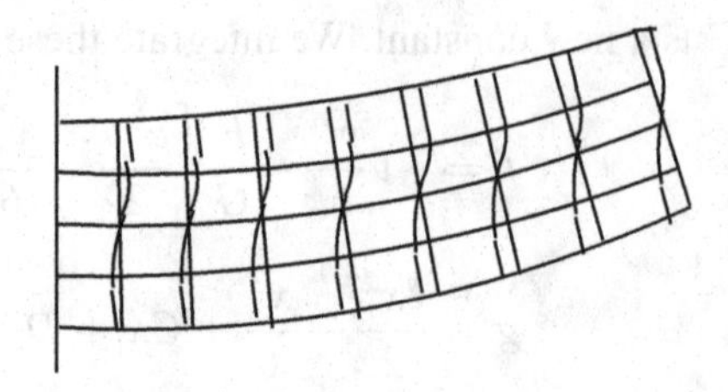

Figure 2.18 Plane sections normal to the neutral axis in the elementary beam theory and the curved sections due to the shear strain.

where our present approximation gives $k = 1.5$. This is attributed to Timoshenko. The value of $k = 6/5$ was obtained by Reissner for static cases and $k = 12/\pi^2$ by Mindlin for dynamic cases. From the parabolic distribution of the shear stress, we know the shear strain is maximum at the neutral axis of the beam and zero at the upper and lower surfaces. Fig. 2.18 shows a comparison of the plane sections normal to the neutral axis in the elementary beam theory and the effect of shear strain in distorting them into a curve.

2.10 Thermal Expansion

Materials expand when heated. In isotropic materials such as poly-crystalline metals, the material expands at an equal strain in all directions. This strain is called the thermal strain and it is expressed as

$$\epsilon_T = \alpha \Delta T, \tag{2.129}$$

where α is the coefficient of linear thermal expansion and ΔT is the change in temperature. These strains contribute to volume change and there is no distortion associated with temperature change. Adding the thermal strains to the mechanical strains in Hooke's law, we get

$$\epsilon_{xx} = \frac{1}{E}[\sigma_{xx} - \nu(\sigma_{yy} + \sigma_{zz})] + \alpha \Delta T, \tag{2.130}$$

$$\epsilon_{yy} = \frac{1}{E}[\sigma_{yy} - \nu(\sigma_{zz} + \sigma_{xx})] + \alpha \Delta T, \tag{2.131}$$

$$\epsilon_{zz} = \frac{1}{E}[\sigma_{zz} - \nu(\sigma_{xx} + \sigma_{yy})] + \alpha \Delta T, \tag{2.132}$$

$$\gamma_{xy} = 2\epsilon_{xy} = \frac{\sigma_{xy}}{G}, \tag{2.133}$$

$$\gamma_{yz} = 2\epsilon_{yz} = \frac{\sigma_{yz}}{G}, \tag{2.134}$$

$$\gamma_{xz} = 2\epsilon_{xz} = \frac{\sigma_{xz}}{G}. \tag{2.135}$$

In many structures, the supports prevent free expansion due to the thermal effect. Then, the mechanical strain and the thermal strain should add up to zero. Thus, thermal effects induce mechanical strains and stresses. These induced stresses are called thermal stresses.

2.10.1 Example: Heated Bar

Consider a bar of length L kept between two rigid walls. If its temperature is raised by ΔT, we have

$$\epsilon_T = \alpha \Delta T. \tag{2.136}$$

The clamped conditions require

$$\epsilon = \epsilon_T + \frac{\sigma}{E} = 0. \tag{2.137}$$

This gives

$$\sigma = -E\alpha\Delta T. \tag{2.138}$$

For a typical aluminum alloy $E = 70$ GPa and $\alpha = 23 \times 10^{-6}$ per degree C. Then the stress generated is 1.61 MPa per degree C. This stress is independent of the length of the bar.

2.11 Saint Venant's Principle

According to this principle, stresses and strains produced by two statically equivalent forces applied on a small area on the surface of an elastic body become indistinguishable far away from the area of application on the surface. Recall that two distributed forces are statically equivalent if they have the same resultant and moment. To illustrate this, consider the two loadings shown in Fig. 2.19(a). The

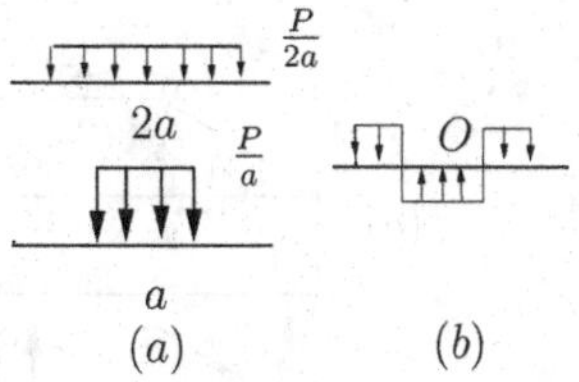

Figure 2.19 Statically equivalent force distributions (*a*) and their difference (*b*).

two distributed loads have the same resultant P and zero moment about the center O. Fig. 2.19(b) shows the difference between the two loads. This distribution has zero resultant and zero moment. St. Venant's principle posits that the stresses and strains produced by the distribution (b) would decrease to zero away from the surface. This statement has been refined by von Mises and Sternberg to show that the solution for the loading at a distance R from the center of loading is proportional to $(2a/R)^{\mu}$, where $\mu > 3$.

We often idealize distributed loads on small areas by concentrated loads. Saint Venant's principle assures that the difference in internal stresses is negligible away from the point of application of the load. We also depend on Saint Venant's principle when we assume uniform state of stress across a tension test specimen at the test section when we do not know the exact loading at the end grips. Saint Venant's principle only applies to static problems; in the dynamic case a disturbance can propagate undiminished far into the interior of an elastic body.

2.12 Anisotropic Materials

When the elastic properties of a material are direction dependent, the material is called anisotropic. This general category can be subdivided into different groups depending on the symmetry property of the material. For example, the simplest subgroup, isotropic materials, have elastic properties invariant under arbitrary rotations about any of the three chosen axes. If the properties are invariant under rotation about one axis, say the z-axis, we have *transverse isotropy*. If they are invariant under reflection about three orthogonal axes, we call it *orthotropy*. In material science, we assume metals are made of a three-dimensional array of atoms forming crystals, which fill the three-dimensional space without any gaps. The classification of anisotropy is based on the shape of the basic crystal. If the material is made from randomly oriented basic crystals or from crystals of different size and orientation, it has no directional properties. These are called polycrystalline materials, and we assume they exhibit isotropic behavior. Many alloys fall into this group.

2.13 Composite Materials

As the name implies, these are composites of different materials. They are nonhomogeneous. Mathematically, the term "nonhomogeneous" implies the material's properties vary from point to point. There are also "functionally graded" materials

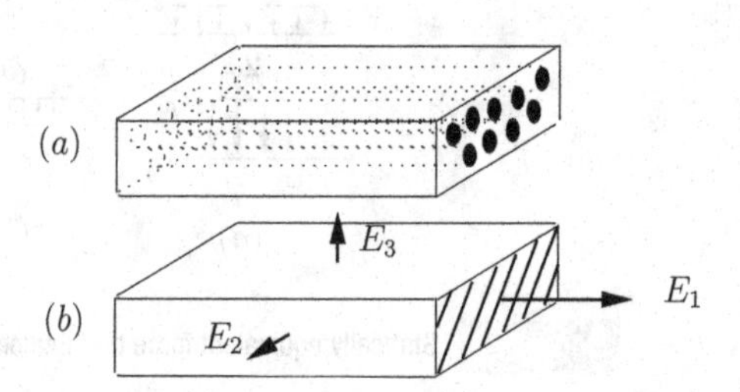

Figure 2.20 Epoxy/fiber composite and an equivalent orthotropic material.

in which the properties do vary continuously. More commonly, we have composite materials such as epoxy/fiber mixtures. In these, the fibers have high values for the Young's modulus and the epoxy's role is to keep the fibers in position without contributing to the stiffness. Fig. 2.20(a) shows the epoxy/fiber arrangement as a nonhomogeneous material. In practice, we may idealize this as a homogeneous material with high stiffness in the fiber direction and low stiffness perpendicular to it. This idealized orthotropic material is shown in Fig. 2.20(b).

The stress–strain relations for orthotropic materials can be written as

$$\begin{aligned}
\epsilon_{xx} &= \frac{1}{E_1}[\sigma_{xx} - \nu_{12}\sigma_{yy} - \nu_{13}\sigma_{zz}],\\
\epsilon_{yy} &= \frac{1}{E_2}[\sigma_{yy} - \nu_{21}\sigma_{xx} - \nu_{23}\sigma_{zz}],\\
\epsilon_{zz} &= \frac{1}{E_3}[\sigma_{zz} - \nu_{31}\sigma_{xx} - \nu_{32}\sigma_{yy}],\\
\gamma_{yz} &= \frac{\sigma_{yz}}{G_{23}}, \quad \gamma_{xz} = \frac{\sigma_{xz}}{G_{31}}, \quad \gamma_{xy} = \frac{\sigma_{xy}}{G_{12}},
\end{aligned} \tag{2.139}$$

where material symmetry requires

$$\frac{\nu_{12}}{E_1} = \frac{\nu_{21}}{E_2}, \quad \frac{\nu_{13}}{E_1} = \frac{\nu_{31}}{E_3}, \quad \frac{\nu_{23}}{E_2} = \frac{\nu_{32}}{E_3}. \tag{2.140}$$

2.13.1 Lamina Properties

The fiber and matrix mixture is processed as a long tape called a lamina with the end purpose of cutting rectangles out of it with the fibers in different directions and stacking them with glue between the layers. The tape is approximately 0.1 mm thick. The tape also called a pre-preg in the trade jargon. The stacked structure is called a laminate. For a basic thin lamina, using the x'-axis along the fiber direction and assuming plane stress condition

$$\{\epsilon'\} = \begin{bmatrix} \frac{1}{E_1} & -\frac{\nu_{12}}{E_1} & 0 \\ -\frac{\nu_{21}}{E_2} & \frac{1}{E_2} & 0 \\ 0 & 0 & \frac{1}{G_{12}} \end{bmatrix} \{\sigma'\}, \tag{2.141}$$

where we have arranged strain and stress tensors as column vectors,

$$\{\epsilon'\} = \begin{Bmatrix} \epsilon_{xx} \\ \epsilon_{yy} \\ \gamma_{xy} \end{Bmatrix}, \quad \{\sigma'\} = \begin{Bmatrix} \sigma_{xx} \\ \sigma_{yy} \\ \sigma_{xy} \end{Bmatrix}, \tag{2.142}$$

and the subscript 1 indicates the fiber direction and 2 indicates the direction perpendicular to it. Typical values for the material constants for three common composites are given in Table 2.1.

Table 2.1. Elastic constants for certain epoxy composites from Jones (1975)

Property	Unit	Glass/epoxy	Boron/epoxy	Graphite/epoxy
E_1	(GPa)	53.78	206.84	206.84
E_2	(GPa)	17.93	20.68	5.17
ν_{12}		0.25	0.3	0.25
G	(GPa)	8.96	6.89	52.59

The inverse of these relations in Eq. (2.141) can be written as

$$\{\sigma'\} = [S]\{\epsilon'\}, \tag{2.143}$$

where the stiffness matrix $[S]$ has the elements

$$S_{11} = \frac{E_1}{1-\nu_{12}\nu_{21}}, \quad S_{12} = \frac{\nu_{12}E_2}{1-\nu_{12}\nu_{21}} = \frac{\nu_{21}E_1}{1-\nu_{12}\nu_{21}}, \tag{2.144}$$

$$S_{22} = \frac{E_2}{1-\nu_{12}\nu_{21}}, \quad S_{33} = G_{12}, \quad S_{13} = S_{23} = 0. \tag{2.145}$$

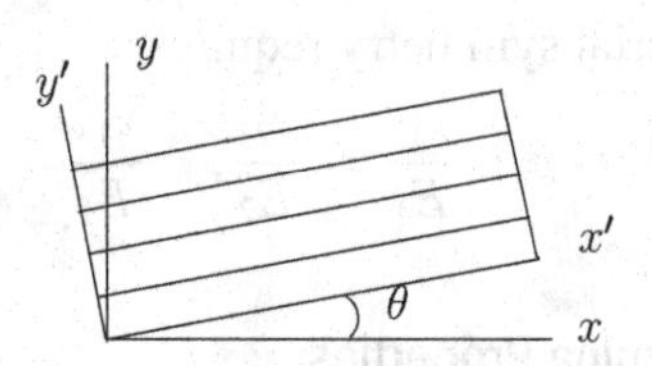

Figure 2.21 Fiber direction x' with a reference frame x, y.

A basic assumption of the analysis of a laminate is that the strains are common to all laminas, but the in-plane stresses may be discontinuous from lamina to lamina. We can transform the stresses and strains to the reference frame x, y using

$$\{\sigma'\} = [Q_\sigma]\{\sigma\}, \quad \{\epsilon'\} = [Q_\epsilon]\{\epsilon'\}, \tag{2.146}$$

where, with $c = \cos\theta$ and $s = \sin\theta$,

$$[Q_\sigma] = \begin{bmatrix} c^2 & s^2 & 2cs \\ s^2 & c^2 & -2cs \\ -cs & cs & c^2 - s^2 \end{bmatrix}, \quad [Q_\epsilon] = \begin{bmatrix} c^2 & s^2 & cs \\ s^2 & c^2 & -cs \\ -2cs & 2cs & c^2 - s^2 \end{bmatrix}, \tag{2.147}$$

where the two transformation matrices are different as we are using the engineering strain γ_{xy} instead of the mathematical strain ϵ_{xy}.

Using Eq. (2.146) in Eq. (2.143), we have

$$[Q_\sigma]\{\sigma\} = [S][Q_\epsilon]\{\epsilon\} \tag{2.148}$$

or

$$\{\sigma\} = [Q_\sigma]^{-1}[S][Q_\epsilon]\{\epsilon\}, \tag{2.149}$$

where

$$[Q_\sigma]^{-1} = \begin{bmatrix} c^2 & s^2 & -2cs \\ s^2 & c^2 & 2cs \\ cs & -cs & c^2 - s^2 \end{bmatrix}. \tag{2.150}$$

The product of the three matrices on the right side is a symmetric matrix

$$[Q_\sigma]^{-1}[S][Q_\epsilon] = [\bar{S}], \tag{2.151}$$

where the elements of $[\bar{S}]$ are:

$$\begin{aligned}
\bar{S}_{11} &= S_{11}c^4 + 2(S_{12} + 2S_{33})c^2s^2 + S_{22}s^4,\\
\bar{S}_{22} &= S_{11}s^4 + 2(S_{12} + 2S_{33})c^2s^2 + S_{22}c^4,\\
\bar{S}_{33} &= (S_{11} + S_{22} - 2S_{12} - 2S_{33})c^2s^2 + S_{33}(c^4 + s^4),\\
\bar{S}_{12} &= (S_{11} + S_{22} - 4S_{33})c^2s^2 + S_{12}(c^4 + s^4),\\
\bar{S}_{13} &= (S_{11} - S_{12} - 2S_{33})c^3s + (S_{12} - S_{22} + 2S_{33})cs^3,\\
\bar{S}_{23} &= (S_{11} - S_{12} - 2S_{33})cs^3 + (S_{12} - S_{22} + 2S_{33})c^3s.
\end{aligned} \tag{2.152}$$

2.13.2 Laminate Properties

Having obtained the stress–strain relations for a lamina with fibers oriented in the θ-direction with respect to an arbitrary coordinate system, x, y, we consider a laminate consisting of a stack of laminae. We restrict ourselves to symmetric stacks, that is, the lamina angles above the midplane of the laminate are a mirror reflection of those below the midplane. Asymmetric laminates may develop bending moments resulting in the curvature of the laminate. Although the strains are constant across the laminate, the stresses may vary from lamina to lamina. In the asymmetric case these stresses are asymmetric, which results in bending moments.

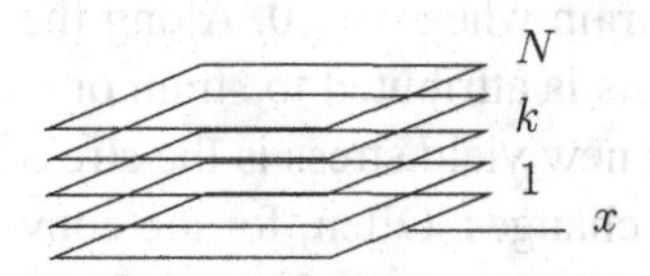

Figure 2.22 A laminate with $2N$ laminae arranged symmetrically about the midplane (only the upper half is shown).

Fig. 2.22 shows a laminate with $2N$ laminae. If we denote an arbitrary lamina above the midplane that has a fiber orientation of θ_k by k, assuming all laminae have the same thickness, the average stresses can be expressed as

$$\bar{\sigma} = \frac{1}{N}\sum_{1}^{N}[S^k]\epsilon, \tag{2.153}$$

where $[S^k]$ is the matrix defined in Eq. (2.152) with $\theta = \theta_k$. Of course, stresses in the individual lamina have to be examined to make sure they are all within the safe range. The stacking sequence for a symmetric laminate is denoted as

$$[\theta_1, \theta_2, \ldots, \theta_N]_S, \tag{2.154}$$

showing the fiber directions starting from the midplane and the subscript S indicates the symmetric lay-up.

2.14 Plasticity in Polycrystalline Metals

In aircraft structures, we avoid plastic deformations as they introduce residual strains. However, a basic knowledge of plasticity models is essential for all mechanics students. A uniaxial test would show the stress proportional to strain for a limited range of strain. This range is called the elastic range. Beyond that, a controlled increase in strain is accompanied by a slight increase in stress. This range is known as the plastic range. The transition point is denoted by the yield strain and the yield stress, σ_Y. A typical stress–strain curve is shown in Fig. 2.23.

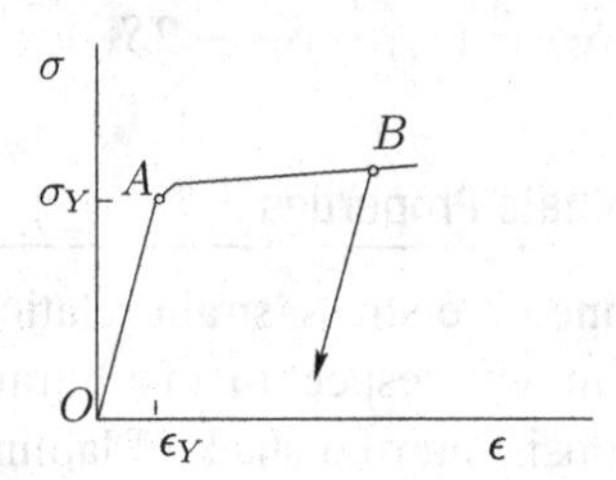

Figure 2.23 A typical uniaxial stress–strain diagram.

In the range, $O-A$, the material is linearly elastic with a slope E, and if we unload the specimen from any point between O and A, the stress–strain curve is unchanged. Beyond the point A, say B, the unloading curve will be linear and parallel to the line OA, but it will be distinct, intersecting the strain axis with a residual strain when $\sigma = 0$. Along the section $A-B$ there is a slight increase in stress. This is attributed to strain or work hardening. If we unload and reload from B, the new yield stress is the stress at B. Thus, due to work hardening, the yield stress changes. Often, for the convenience of simplifying calculations, we approximate the stress–strain curve by the elastic–perfectly plastic model shown in Fig. 2.24, which neglects work hardening.

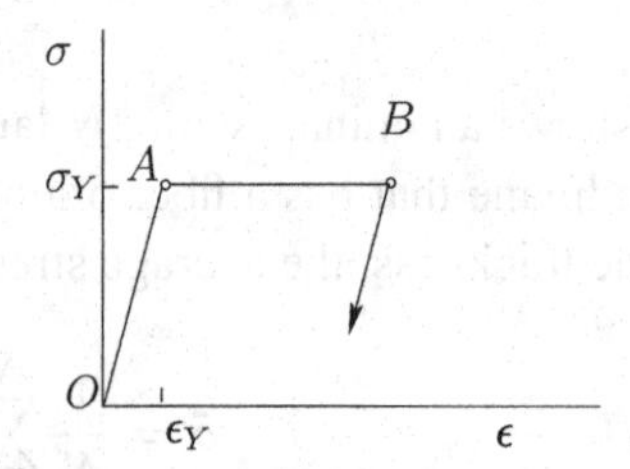

Figure 2.24 The elastic–perfectly plastic model.

A more complex model that approximates the work hardening is the elastic–linear hardening model shown in Fig. 2.25.

Generally, when unloading from a point B if we go to the compressive side, we find compressive yielding. The absolute value of the compressive yield stress is found to be less than the tensile yield stress. This phenomenon is called the Bauschinger effect. In many metals a clear yield point demarcating elastic and

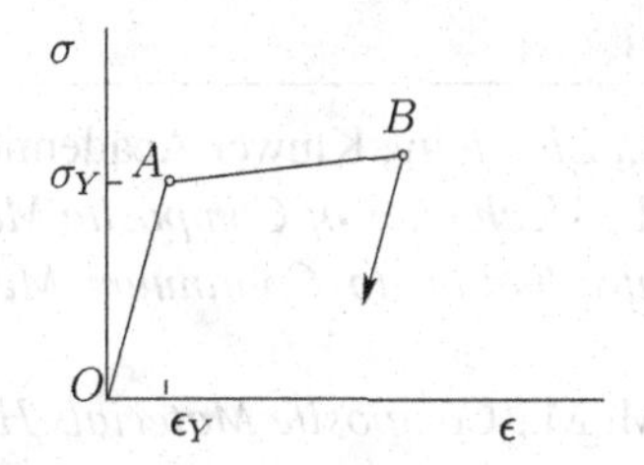

Figure 2.25 The elastic-linear hardening model.

plastic behavior is difficult to establish. Then, a line parallel to the linear elastic loading curve is drawn through 0.2% strain and the stress at its intersection with the stress–strain curve is designated as the yield stress. Instead of the two formulas, one for stresses below the yield stress and one for stresses above the yield stress, a unified formula was introduced by Ramberg and Osgood for aluminum alloys and other metals in the form

$$\epsilon = \frac{\sigma}{E} + K\left(\frac{\sigma}{E}\right)^n. \tag{2.155}$$

The two additional constants K and n are obtained from experiments.

2.15 Commonly Used Aircraft Materials

In aircraft construction, critical structural components such as brackets, bulkheads, spars, and landing gears are usually made of steel, which has a Young's modulus of 200 GPa and Poisson's ratio of 0.32. Depending on the added alloying components the yield stress and ultimate strength may vary around 1500 MPa and 1800 MPa, respectively. This high stiffness and strength is obtained at the expense of a high density of 7.8 gram per cubic centimeter.

Most of a conventional aircraft is made of aluminum, which has a Young's modulus of 71 GPa and Poisson's ratio of 0.33. The yield stress varies from 324 MPa (2024-T3 heat-treated alloy) to 490 MPa (7075-T6). It has density of 2.78 g/cm^3. Another expensive material used for its high strength and low density is titanium. This is particularly suitable for high-temperature applications such as engine components. A major drawback of metals is their strength deterioration under cyclic loading, which is known as fatigue.

These days more and more noncritical components are made of composite materials. The basic epoxy/fiber arrangements of orthotropic layers is stacked at different angles to form laminated composites. These have exceptional stiffness in the desired directions, low density, and high durability under cyclic loading. The carbon fibers have, approximately, $E = 240$ GPa and glass fibers have $E = 80$ GPa, with densities of 1.8 g/cm^3 and 2.5 g/cm^3, respectively. These materials do not exhibit fatigue. They are sensitive to humidity and other environmental conditions. Also, their failure is usually abrupt without any warning signs.

FURTHER READING

Barber, J. R., *Elasticity*, Kluwer Academic (2002).
Jones, R. M., *Mechanics of Composite Materials*, McGraw-Hill (1975).
Nair, S., *Introduction to Continuum Mechanics*, Cambridge University Press (2009).
Schwartz, M. M., *Composite Materials Handbook*, McGraw-Hill (1983).
Timoshenko, S. P. and Goodier, J. N., *Theory of Elasticity*, McGraw-Hill (1987).
Zhang, S. and Zhao, D. (editors), *Aerospace Materials Handbook*, CRC Press (2012).

EXERCISES

2.1 The stress matrix at a point in a material is given by

$$[\sigma] = \begin{bmatrix} 4 & 2 & 0 \\ 2 & 3 & 1 \\ 0 & 1 & 5 \end{bmatrix}.$$

Obtain the traction vector on the plane

$$2x + y + 2z = 0.$$

Compute the normal stress and the shear stress on this plane.

2.2 For the stress matrix expressed using the principal coordinates,

$$[\sigma] = \begin{bmatrix} \sigma_1 & 0 & 0 \\ 0 & \sigma_2 & 0 \\ 0 & 0 & \sigma_3 \end{bmatrix},$$

obtain the normal stress on one of the octahedral planes in the first octant. The normal to an octahedral plane makes equal angles with the three coordinate axes. Also, compute the shear stress on this plane.

2.3 Assume a cantilever beam occupies the region, $0 < x < \ell$, $-c < y < c$, $-b < z < b$ and a state of plane stress exists in the x, y-plane with the only body force component $f_y = -\rho g$ where ρ is the density and g is the gravitational constant. See Fig. 2.26.

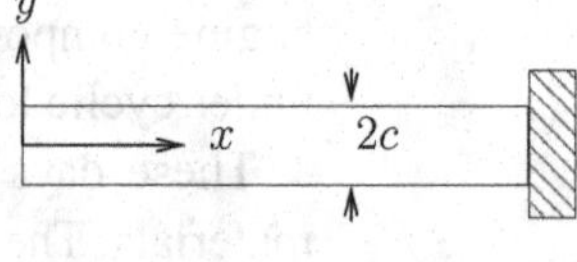

Figure 2.26 Bending of a cantilever beam under its own weight.

If

$$\sigma_{xx} = -\frac{My}{I}, \quad I = \frac{4}{3}bc^3,$$

where, under its own weight, the beam has

$$M = -2\rho g b c x^2.$$

Using the equilibrium equations, obtain the shear stress σ_{xy} and the normal stress σ_{yy} such that are traction vectors on the planes $y = \pm c$.

2.4 At a point the state of stress is

$$[\sigma] = \begin{bmatrix} \sigma & 2 & 0 \\ 2 & 3 & 1 \\ 0 & 1 & 5 \end{bmatrix},$$

where σ has to be found. Knowing that there is a plane through this point on which the traction vector is zero, compute the value of the unknown, σ.

2.5 In the case of plane stress in the x, y-plane, show that the transformation law for the components of the stress tensor in a new x', y' coordinate system obeys

$$\sigma' = [Q][\sigma[Q]^T,$$

where $[Q]$ is the rotation matrix

$$[Q] = \begin{bmatrix} \cos\theta & \sin\theta \\ -\sin\theta & \cos\theta \end{bmatrix},$$

where θ is the angle from the x-axis to the x'-axis.

2.6 Transform the components of the tensor

$$[\sigma] = \begin{bmatrix} 10 & 3 \\ 3 & 2 \end{bmatrix}$$

to a new system rotated counterclockwise by 30° from the old system.

2.7 Using the method of focus draw Mohr's circle and show the properly oriented principal planes, maximum shear stress planes, and the x and y planes with the correct normal and shear stresses on them for the state of stress in Fig. 2.27.

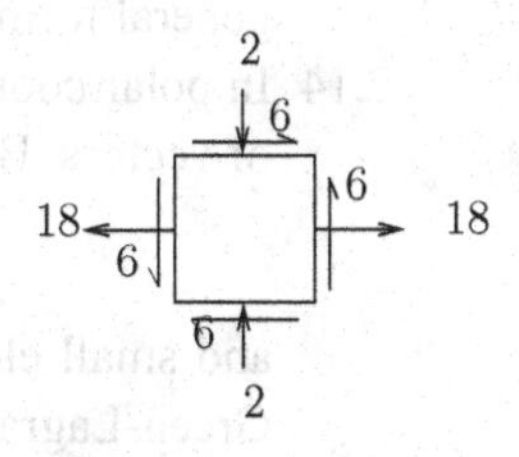

Figure 2.27 Stresses on an element.

2.8 For the stress matrix

$$[\sigma] = \begin{bmatrix} x^2y & (a^2 - y^2)x & 0 \\ (a^2 - y^2)x & (y^3 - 3a^2y)/3 & 0 \\ 0 & 0 & 2az^2 \end{bmatrix},$$

where a is a constant, obtain the components of the body force vector necessary for equilibrium.

2.9 The displacement components in a body are given by

$$u = Axy, \quad v = A(x^2 - y^2), \quad w = 0.$$

Obtain the components of the Green-Lagrange strain tensor.

2.10 Using the method of focus draw Mohr's circle and show the properly oriented principal planes, maximum shear stress planes, and the x and y planes with the correct normal and shear stresses on them for the state of stress in Fig. 2.28.

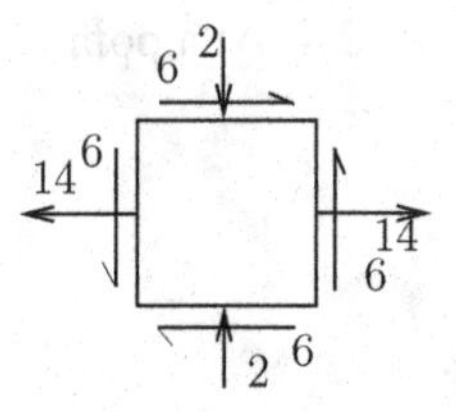

Figure 2.28 Stresses on an element.

2.11 Draw Mohr's circle for the state of stress shown in Fig. 2.29 and show the principal planes and the maximum shear planes, using the method of focus.

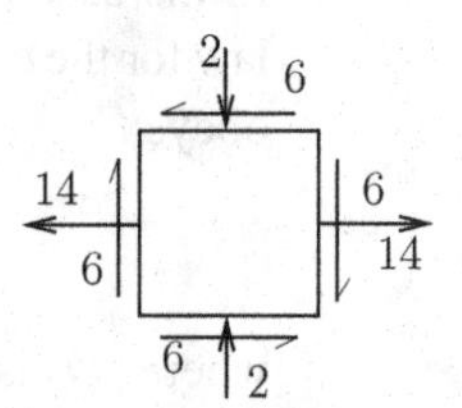

Figure 2.29 Stresses on an element.

2.12 A strain rosette has three resistors, R_1, R_2, and R_3, arranged at 120° apart, with R_1 being along the x-axis. If they register strains, 3×10^{-6}, 5×10^{-6}, and 8×10^{-6}, respectively, obtain the strain matrix.

2.13 The three components of the linear strain in a plane strain case are given by

$$\epsilon_{xx} = Ay^2, \quad \epsilon_{yy} = Ax^2, \quad \gamma_{xy} = Cxy.$$

Using the compatibility equations, express C in terms of A. Obtain the general form of displacements for this case.

2.14 In polar coordinates, we use unit vectors $\boldsymbol{i}_r$ and $\boldsymbol{i}_\theta$ to express components of vectors. Both of these depend on the coordinate θ. Using

$$\boldsymbol{u} = u(r, \theta)\boldsymbol{i}_r + v(r, \theta)\boldsymbol{i}_\theta$$

and small elements directed along r and θ obtain the components of the Green-Lagrange strain in polar coordinates. Also find its linear version.

2.15 What is the compatibility equation relating ϵ_{rr}, $\epsilon_{\theta\theta}$ and $\gamma_{r\theta}$?

2.16 A thin plate in the x, y-plane has a square shape with sides 3 cm. It undergoes a unform strain of

$$\epsilon_{xx} = 3 \times 10^{-6}, \quad \epsilon_{yy} = 5 \times 10^{-6}, \quad \epsilon_{xy} = -4 \times 10^{-6}.$$

If the point $x = 0$, $y = 0$ has no displacement and the edge $x = 0$ has no rotation, obtain the displacement distribution inside the plate. Specifically, what are u and v at (2, 1)?

2.17 The Airy stress function, $\phi(x, y)$, is defined through the relations

$$\frac{\partial^2 \phi}{\partial y^2} = \sigma_{xx}, \quad \frac{\partial^2 \phi}{\partial x^2} = \sigma_{yy}, \quad \frac{\partial^2 \phi}{\partial x \partial y} = -\sigma_{xy}.$$

Show that the equilibrium equations are satisfied if the body force vector is zero and

$$\sigma_{xz} = \sigma_{yz} = 0, \quad \sigma_{zz} = \sigma_{zz}(x, y).$$

Express the compatibility equation in terms of ϕ for the plane stress and the plane strain cases using Hooke's law. Simplify your result using the notation for the Laplace operator

$$\nabla^2 = \frac{\partial^2}{\partial x^2} + \frac{\partial^2}{\partial y^2}.$$

2.18 In polar coordinates we have normal stresses σ_{rr} along the r-direction, $\sigma_{\theta\theta}$ along the θ-direction, and shear stress $\sigma_{r\theta}$. Using the unit vectors $\boldsymbol{i}_r$ and $\boldsymbol{i}_\theta$ and the body force components f_r and f_θ obtain the equilibrium equations for the plane stress case.

2.19 A glass/epoxy composite lamina has $E_1 = 53.75$ GPa, $E_2 = 17.92$ GPa, $G_{12} = 8.96$ GPa, and $\nu_{12} = 0.25$. Plot the stiffnesses S_{11}/G_{12}, S_{22}/G_{12}, and S_{13}/G_{12} as functions of θ.

2.20 A composite laminate is constructed using glass/epoxy laminae with the properties $E_1 = 53.75$ GPa, $E_2 = 17.92$ GPa, $G_{12} = 8.96$ GPa, and $\nu_{12} = 0.25$. The stacking sequence is given as

$$[0°, 45°, -45°, 0°]_S.$$

Obtain the average stiffness matrix $[\bar{S}]$. For the given strain components, $\epsilon_{xx} = 4 \times 10^{-6}$, $\epsilon_{yy} = 2 \times 10^{-6}$, and $\gamma_{xy} = 3 \times 10^{-6}$, compute the average stress components in the laminate.

3 Energy Methods

Energy methods provide an approach to obtain relations between forces and displacements in structures through the minimization of a scalar function. These methods can be classified as (a) force methods and (b) displacement methods. As we know, energy is a scalar quantity and it is simpler to deal with compared to vector quantities. These methods can be used to find deflections under applied loads or vice versa. They can also be useful in approximating solutions. We begin with a simple presentation of virtual work theories, which form the starting point for energy methods.

When a load P moves by a small distance du, the incremental work done is given by

$$dW = Pdu.$$

This shown in Fig. 3.1. The quantity udP is called the complementary work, dW^*. When the displacements are virtual, that is, fictitious, we denote them by

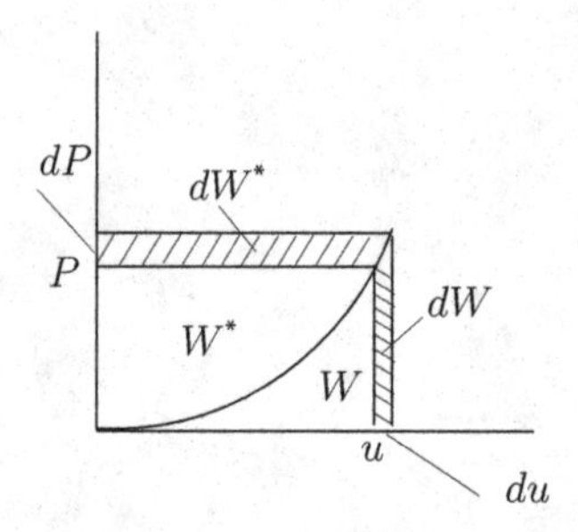

Figure 3.1 Work and complementary work.

δu. Similarly virtual forces are denoted by δP. Work and complementary work involving these virtual quantities are, similarly, written as δW and δW^*.

In Fig. 3.2(a) we have a particle subjected to a given system of forces undergoing a virtual displacement $\boldsymbol{\delta u}$. In a complicated structure, virtual displacements have to be selected to maintain compatibility. For example, in the case of a beam these assumed shapes should not introduce any hinges or breaks. As mentioned earlier, we denote virtual quantities with a δ in front. The virtual work done by the system of forces is

$$\delta W = \boldsymbol{P}_1 \bullet \boldsymbol{\delta u} + \boldsymbol{P}_2 \bullet \boldsymbol{\delta u} + \boldsymbol{P}_3 \bullet \boldsymbol{\delta u}. \tag{3.1}$$

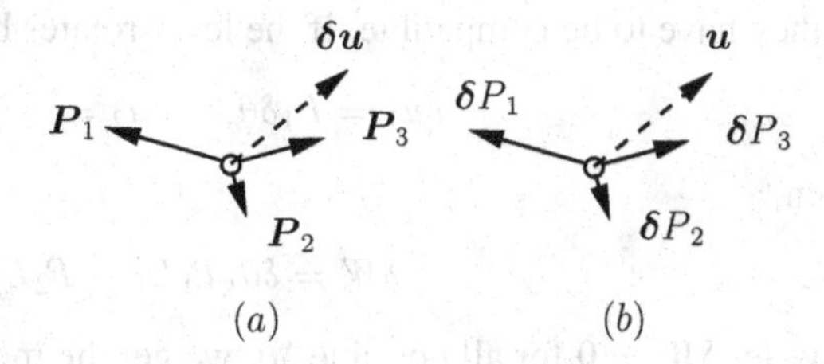

Figure 3.2 A particle subjected to actual forces moving by a virtual displacement $\delta \boldsymbol{u}$ in (a) and a particle subjected to virtual forces moving by a real displacement $\boldsymbol{u}$ in (b).

This can written as

$$\delta W = (\boldsymbol{P}_1 + \boldsymbol{P}_2 + \boldsymbol{P}_3) \bullet \delta \boldsymbol{u}. \tag{3.2}$$

If $\delta W = 0$ for all possible choices of $\delta \boldsymbol{u}$, we conclude

$$\boldsymbol{P}_1 + \boldsymbol{P}_2 + \boldsymbol{P}_3 = \boldsymbol{0}, \tag{3.3}$$

which is the requirement for equilibrium. Thus, equilibrium of a system of forces is implied by the condition that the virtual work done on the body is zero.

In Fig. 3.2(b) we have a system of virtual forces in equilibrium.

$$\delta \boldsymbol{P}_1 + \delta \boldsymbol{P}_2 + \delta \boldsymbol{P}_3 = \boldsymbol{0}.$$

Under a real displacement $\boldsymbol{u}$ we see the virtual complementary work done is

$$\delta W^* = \delta \boldsymbol{P}_1 \bullet \boldsymbol{u} + \delta \boldsymbol{P}_2 \bullet \boldsymbol{u} + \delta \boldsymbol{P}_3 \bullet \boldsymbol{u}, \tag{3.4}$$

$$= (\delta \boldsymbol{P}_1 + \delta \boldsymbol{P}_2 + \delta \boldsymbol{P}_3) \bullet \boldsymbol{u}, \tag{3.5}$$

$$= 0. \tag{3.6}$$

With a single displacement $\boldsymbol{u}$, the virtual force principle here does not illustrate the enforcement of compatibility.

Let us apply these principles to the lever shown in Fig. 3.3. In (a), the virtual displacements δu_1 and δu_2 have to be imposed without breaking the lever, that

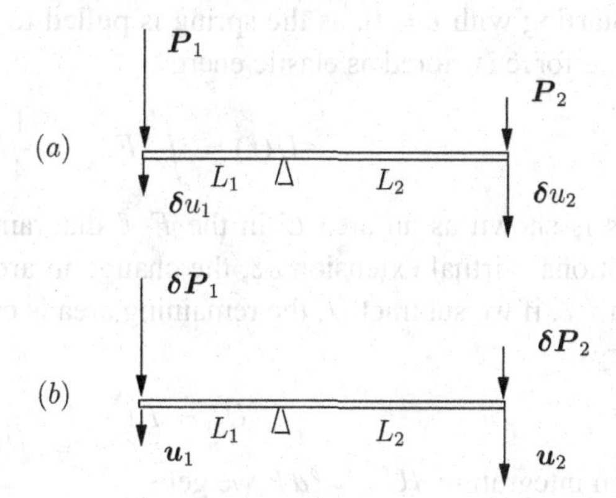

Figure 3.3 A lever subjected to actual forces moving by virtual displacements in (a) and a lever subjected to virtual forces moving by actual displacements in (b).

is, they have to be compatible. If the lever rotates by a virtual angle $\delta\theta$,

$$\delta u_1 = L_1\delta\theta, \quad \delta u_2 = -L_2\delta\theta.$$

Then,

$$\delta W = \delta\theta(P_1L_1 - P_2L_2).$$

If we let $\delta W = 0$ for all possible $\delta\theta$, we get the moment equilibrium,

$$P_1L_1 - P_2L_2 = 0.$$

In (b), the virtual forces have to be selected meeting the requirement of equilibrium. Thus,

$$\delta P_1L_1 - \delta P_2L_2 = 0.$$

The complementary virtual work becomes

$$\delta W^* = \delta P_1 u_1 + \delta P_2 u_2 = \delta P_1 \left[u_1 + \frac{L_1}{L_2} u_2 \right] = 0.$$

From this we find the compatible displacements due to a rotation θ given by

$$\theta = \frac{u_1}{L_1} = -\frac{u_2}{L_2}.$$

This simple example illustrates the complementary roles of equilibrium of forces and compatibility of displacements.

3.1 Strain Energy and Complementary Strain Energy

Once we are familiar with the virtual work principles for rigid bodies, we proceed to elastic bodies. Elastic bodies have stored energies and the virtual work done on them will alter the stored energy. Thus $\delta W \neq 0$. Let us begin with a nonlinear spring, which has the force–displacement relation

$$F = k\ell^3, \tag{3.7}$$

where F is the internal force and ℓ is the extension of the spring.

Starting with $\ell = 0$, as the spring is pulled to an extension ℓ, the work done by the force is stored as elastic energy,

$$U(\ell) = \int_0^\ell F d\ell = \frac{1}{4}k\ell^4.$$

This is shown as an area U in the F–ℓ diagram in Fig. 3.4. If we impose an additional virtual extension $\delta\ell$, the change in area is δU. From the rectangular area $F\ell$, if we subtract U, the remaining area is called the complementary strain energy,

$$U^* = F\ell - U(\ell).$$

From integrating $dU^* = \ell dF$ we get

$$U^*(F) = \int_0^F \ell dF = \frac{1}{k^{1/3}} \int_0^F F^{1/3} dF = \frac{3}{4k^{1/3}} F^{4/3}.$$

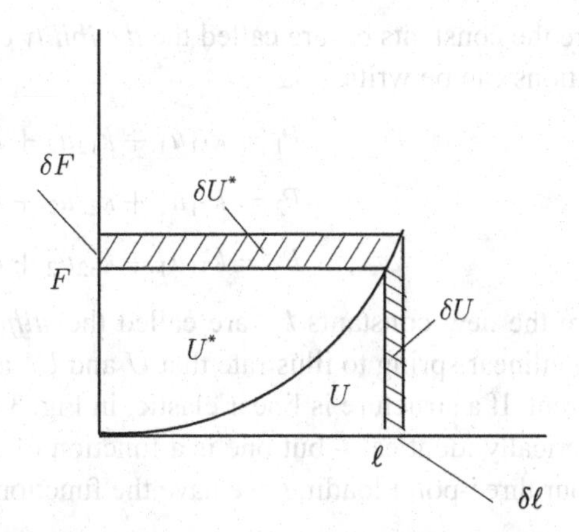

Figure 3.4 Strain energy and complementary strain energy.

The relation between the functions U and U^* is analogous to that between the internal energy and enthalpy in thermodynamics where the variables are pressure and specific volume instead of force and extension.

We distinguish between internal quantities and externally applied quantities. If this spring is under an applied external load P and we give a virtual displacement δu to the tip of the spring, compatibility requires that the virtual extension $\delta\ell = \delta u$. We get, by equating virtual work to the change in strain energy,

$$P\delta u = \delta U, \quad P\delta u = k\ell^3 \delta u, \quad P = k\ell^3, \quad P = F,$$

which is the equilibrium equation relating the internal and external loads. If it is stretched to a distance u under a load P then equilibrium requires that the internal force $F = P$ and if we apply a virtual force δP at its end we have $\delta F = \delta P$. Using virtual complementary work, we find

$$u\delta P = \delta U^*, \quad u\delta P = \frac{1}{k^{1/3}} F^{1/3} \delta F = \frac{1}{k^{1/3}} F^{1/3} \delta P, \quad u = \frac{1}{k^{1/3}} F^{1/3}, \quad u = \ell,$$

which is the compatibility of displacements.

3.1.1 Flexibility and Stiffness Coefficient

Let us consider a truss with applied forces P_1, P_2, and P_3 at three node points. The displacements under these loads oriented in the direction of the loads will be denoted by u_1, u_1, and u_3. We restrict ourselves to three loads, but extending them to an arbitrary number of loads is straight forward. In a linear elastic material the system of forces and displacements are linearly related in the form

$$\begin{aligned} u_1 &= c_{11}P_1 + c_{12}P_2 + c_{13}P_3, \\ u_2 &= c_{21}P_1 + c_{22}P_2 + c_{23}P_3, \\ u_3 &= c_{31}P_1 + c_{32}P_2 + c_{33}P_3, \end{aligned} \tag{3.8}$$

where the constants c_{ij} are called the *flexibility coefficients*. The inverse of these equations can be written as

$$\begin{aligned} P_1 &= k_{11}u_1 + k_{12}u_2 + k_{13}u_3, \\ P_2 &= k_{21}u_1 + k_{22}u_2 + k_{23}u_3, \\ P_3 &= k_{31}u_1 + k_{32}u_2 + k_{33}u_3, \end{aligned} \tag{3.9}$$

where the new constants k_{ij} are called the *stiffness coefficients*. We have used the nonlinear spring to illustrate that U and U^* are numerically and functionally different. If a structure is linear elastic, in Fig. 3.4 the areas of U and U^* will be numerically identical – but one is a function of ℓ and the other a function of F. For our three-point loading, we have the functional dependence,

$$U = U(u_1, u_2, u_3), \quad U^* = U^*(P_1, P_2, P_3). \tag{3.10}$$

If we apply a virtual force δP_1 at point 1, from the virtual complementary energy principle,

$$u_1 \delta P_1 = \delta U^* = \frac{\partial U^*}{\partial P_1}\delta P_1, \quad u_1 = \frac{\partial U^*}{\partial P_1}. \tag{3.11}$$

Generalizing this to any node i,

$$u_i = \frac{\partial U^*}{\partial P_i}. \tag{3.12}$$

This is the **first theorem of Castigliano**.

By applying a virtual displacement δu_1 on node 1, we get

$$P_1 \delta u_1 = \delta U = \frac{\partial U}{\partial u_1}\delta u_1, \quad P_1 = \frac{\partial U}{\partial u_1}. \tag{3.13}$$

From this

$$P_i = \frac{\partial U}{\partial u_i}, \tag{3.14}$$

which is the **second theorem of Castigliano**.

Castigliano's first theorem combined with Eq. (3.8) gives

$$\begin{aligned} \frac{\partial U^*}{\partial P_1} &= c_{11}P_1 + c_{12}P_2 + c_{13}P_3, \\ \frac{\partial U^*}{\partial P_2} &= c_{21}P_1 + c_{22}P_2 + c_{23}P_3. \end{aligned}$$

Differentiating the first equation with P_2 and the second by P_1 and noting that in the mixed derivative of U^* the order of differentiation commutes, we get

$$c_{12} = c_{21}.$$

We can generalize this to show that the flexibility matrix $[c_{ij}]$ and the stiffness matrix $[k_{ij}]$ are symmetric. In the literature, the first and the second theorems are often interchanged.

It is worth noting that we may add moments to the list of applied forces provided we use the angle of rotation of the moments in the displacement list, as work done by a moment is the product of the moment and its angle of rotation.

Before we apply these principles to elastic structures we have to get expressions for the stored energy.

In the linear elastic problems, the statement "stored elastic energy is equal to the work done by applied loads" is known as Clapeyron's theorem.

3.2 Strain Energy Density

If we consider an element of length Δx with a cross-sectional area $\Delta A = \Delta y \Delta z$, it will stretch to a length of $\delta x(1 + \epsilon_{xx})$ under a normal stress σ_{xx}. The work done by the force $\sigma_{xx}\Delta A$ in moving through a distance $\Delta x \epsilon_{xx}$ is the energy stored,

$$U_T = \frac{1}{2}\sigma_{xx}\epsilon_{xx}\Delta V, \quad \Delta V = \Delta x \Delta y \Delta z,$$

where we use the subscript T to indicate the total energy. We define the strain energy density U as the energy stored per unit volume, that is,

$$U = \frac{1}{2}\sigma_{xx}\epsilon_{xx}. \tag{3.15}$$

The energy stored from the work done by shear stress σ_{xy} can be assessed from Fig. 3.5.

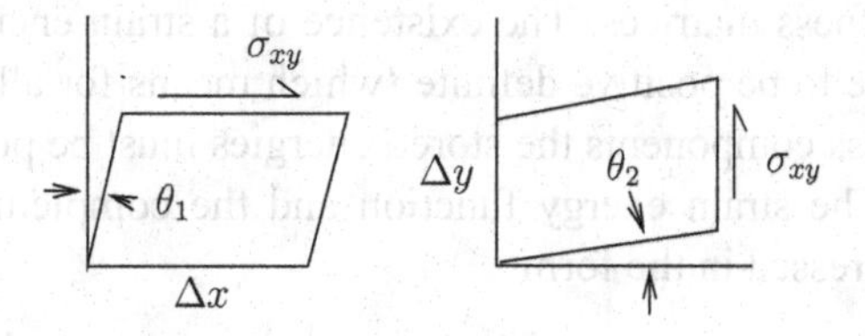

Figure 3.5 Work done by shear stresses.

Here

$$\begin{aligned} U_T &= \frac{1}{2}[(\sigma_{xy}\Delta x \Delta z)(\Delta y \theta_1) + (\sigma_{xy}\Delta y \Delta z)(\Delta x \theta_2)], \\ &= \frac{1}{2}\sigma_{xy}\gamma_{xy}\Delta V. \end{aligned}$$

The energy density is

$$U = \frac{1}{2}\sigma_{xy}\gamma_{xy}. \tag{3.16}$$

For the three-dimensional case, by adding the work done by all the stresses, we get

$$U = \frac{1}{2}[\sigma_{xx}\epsilon_{xx} + \sigma_{yy}\epsilon_{yy} + \sigma_{zz}\epsilon_{zz} + \sigma_{xy}\gamma_{xy} + \sigma_{yz}\gamma_{yz} + \sigma_{xz}\gamma_{xz}]. \tag{3.17}$$

3.2.1 Voigt Notation

Instead of dealing with stress and strain as square matrices, Voigt introduced the following notation:

$$\{\sigma\} = \begin{Bmatrix} \sigma_1 \\ \sigma_2 \\ \sigma_3 \\ \sigma_4 \\ \sigma_5 \\ \sigma_6 \end{Bmatrix} = \begin{Bmatrix} \sigma_{xx} \\ \sigma_{yy} \\ \sigma_{zz} \\ \sigma_{yz} \\ \sigma_{xz} \\ \sigma_{xy} \end{Bmatrix}, \quad \{\epsilon\} = \begin{Bmatrix} \epsilon_1 \\ \epsilon_2 \\ \epsilon_3 \\ \epsilon_4 \\ \epsilon_5 \\ \epsilon_6 \end{Bmatrix} = \begin{Bmatrix} \epsilon_{xx} \\ \epsilon_{yy} \\ \epsilon_{zz} \\ \gamma_{yz} \\ \gamma_{xz} \\ \gamma_{xy} \end{Bmatrix}. \tag{3.18}$$

In this notation the generalized Hooke's law for isotropic materials can be written as

$$\{\epsilon\} = \frac{1}{E}\begin{bmatrix} 1 & -\nu & -\nu & 0 & 0 & 0 \\ -\nu & 1 & -\nu & 0 & 0 & 0 \\ -\nu & -\nu & 1 & 0 & 0 & 0 \\ 0 & 0 & 0 & 2(1+\nu) & 0 & 0 \\ 0 & 0 & 0 & 0 & 2(1+\nu) & 0 \\ 0 & 0 & 0 & 0 & 0 & 2(1+\nu) \end{bmatrix}\{\sigma\}. \tag{3.19}$$

For anisotropic materials we have

$$\{\epsilon\} = [C]\{\sigma\}, \quad \{\sigma\} = [S]\{\epsilon\}, \tag{3.20}$$

where the symmetric matrices $[C]$ and $[S]$ are called the elastic compliance and stiffness matrices. The existence of a strain energy requires that these matrices have to be positive definite (which means for all nonzero choices for strain and stress components the stored energies must be positive).

The strain energy function and the complementary energy function can be expressed in the form

$$U = \frac{1}{2}\{\epsilon\}^T[C]\{\epsilon\}, \quad U^* = \frac{1}{2}\{\sigma\}^T[S]\{\sigma\}. \tag{3.21}$$

3.2.2 Uniform Bar Subjected to an Axial Tension

For an axially loaded member with cross-sectional area A and Young's modulus E,

$$\sigma_{xx} = E\epsilon_{xx}, \quad U = \frac{1}{2}E\epsilon_{xx}^2.$$

Using $\epsilon_{xx} = du/dx$, the energy per unit length of this bar is

$$U = \frac{1}{2}EA(u')^2, \quad u' \equiv \frac{du}{dx}.$$

If the internal force is F, $\sigma_{xx} = F/A$. The complementary energy per unit volume is

$$U^* = \frac{1}{2E}\sigma_{xx}^2 = \frac{F^2}{2EA^2}.$$

Per unit length of the bar, multiplying the given density by A, we get

$$U^* = \frac{F^2}{2EA}.$$

For a bar of length L with uniform cross-section and constant value for E, under a tensile load F,

$$\frac{du}{dx} = \frac{\ell}{L}, \tag{3.22}$$

where ℓ is its extension (change in length), and the total energies are

$$U_T = \frac{1}{2}EA(\ell/L)^2 L = \frac{EA\ell^2}{2L}, \quad U_T^* = \frac{1}{2}\frac{F^2}{EA}L = \frac{F^2 L}{2EA}. \tag{3.23}$$

The virtual changes are given by

$$\delta U_T = \frac{EA\ell\delta\ell}{L}, \quad \delta U_T^* = \frac{F\delta F L}{EA}. \tag{3.24}$$

3.2.3 Circular Shaft Subjected to a Torque

Using a cylindrical coordinate system, (r, θ, z), the nonzero stress is $\tau = \sigma_{z\theta}$ and strain is $\gamma = \gamma_{z\theta}$, which are related as

$$\gamma = \frac{\tau}{G}. \tag{3.25}$$

For a uniform shaft of radius c and length L, if the relative rotation of one end of the shaft with respect to the other is θ,

$$\gamma = \frac{\theta r}{L}. \tag{3.26}$$

The energy density $G\gamma^2/2$ can be integrated across the cross-section to find the energy per unit length as

$$U = \frac{1}{2}\int_0^c G\gamma^2 2\pi r dr = \frac{1}{2}GJ\alpha^2, \quad J = \frac{\pi c^4}{2}, \quad \alpha = \frac{\theta}{L}, \tag{3.27}$$

where α is the rate of twist and J is the polar second moment of the area.

Using $\gamma = \alpha r$ from Eq. (3.26), $\tau = G\alpha r$, and the internal torques, T is obtained as

$$T = \int_0^c \tau r^2 2\pi dr = GJ\alpha. \tag{3.28}$$

We may convert the strain energy to the complementary energy by substituting for α in terms of T, to have

$$U^* = \frac{1}{2}\frac{T^2}{GJ}. \tag{3.29}$$

For a shaft of length L with uniform properties we get

$$U_T = \frac{GJ\theta^2}{2L}, \quad U_T^* = \frac{T^2 L}{2GJ}, \tag{3.30}$$

and their virtual changes are

$$\delta U_T = \frac{GJ\theta\delta\theta}{L}, \quad U_T^* = \frac{T\delta TL}{GJ}. \tag{3.31}$$

3.2.4 Beam Subjected to a Moment

Consider a beam of width b and height $2c$. Under elementary beam theory assumptions, $\sigma = \sigma_{xx}$ and $\epsilon = \epsilon_{xx}$ are given by

$$\sigma = \frac{My}{I}, \quad I = \frac{2}{3}bc^3, \quad \epsilon = \frac{y}{R}, \quad \frac{1}{R} = \frac{d^2v}{dx^2}. \tag{3.32}$$

The energy density per unit volume, $E\epsilon^2/2$, can be integrated across the cross-section to get the energy density per unit length,

$$U = \frac{1}{2}\int_{-c}^{c} E\epsilon^2 bdy = \frac{1}{2}EI(v'')^2. \tag{3.33}$$

$$U^* = \frac{1}{2E}\int_{c}^{c} \sigma^2 bdy = \frac{1}{2}\frac{M^2}{EI}. \tag{3.34}$$

The changes in energy densities are given by

$$\delta U = EI(v'')(\delta v)'', \quad \delta U^* = \frac{M\delta M}{EI}. \tag{3.35}$$

3.3 Two-bar Truss

Consider the two-bar truss shown in Fig. 3.6. The two bars: Bar 1, BC, and bar 2, AC, have the same values for EA. We assume the applied loads P and Q are

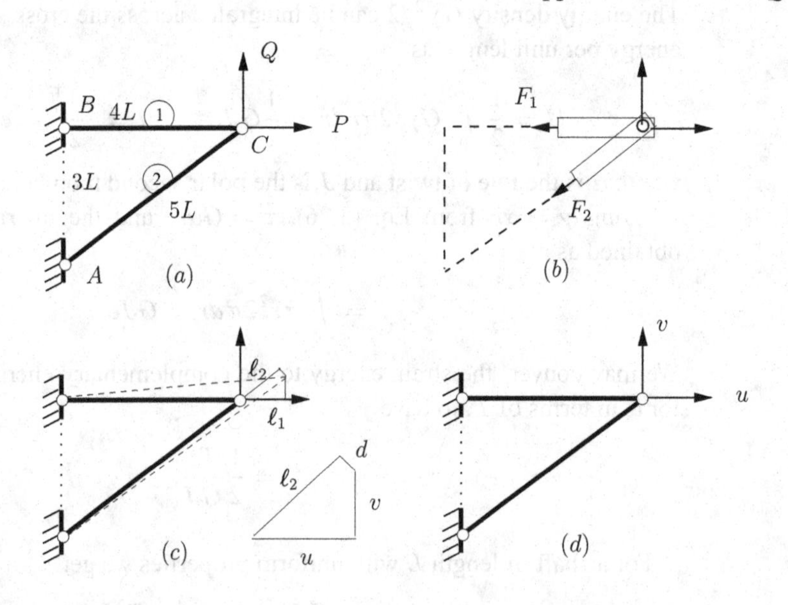

Figure 3.6 Two-bar truss: (a) geometry and loads; (b) free-body diagram to find internal forces; (c) extensions of individual members and compatible reconnection; and (d) as a prescribed displacement problem.

known. The displacements along P and Q are u and v, respectively. Taking the moment about A, the internal forces can be obtained as follows:

$$F_1(3L) = P(3L) - Q(4L), \quad F_1 = P - \frac{4}{3}Q.$$

Vertical Forces at the hinge C give

$$F_2(3/5) - Q = 0, \quad F_2 = 5Q/3.$$

3.3.1 Solution Without Using Energy

In order to appreciate the simplicity of energy methods, it is worthwhile to solve this simple problem directly. Having obtained the member forces, we compute the changes in the lengths of the two members as

$$\ell_1 = \frac{F_1 L_1}{EA}, \quad \ell_2 = \frac{F_2 L_2}{EA},$$

where

$$L_1 = 4L, \quad L_2 = 5L.$$

To find the final location of the joint C, we reconnect the two deformed bars by rotating them. When the extensions are small the displacement contribution from rotation is assumed to be perpendicular to the original position of a member. In this process, we see the horizontal deflection of C,

$$u = \ell_1 = \frac{4F_1 L}{EA} = \frac{L}{EA}\left[4P - \frac{16}{3}Q\right]. \tag{3.36}$$

To find the vertical deflection v, we use the four-sided figure amplified in Fig. 3.6(c). We have

$$\sin\theta = 3/5, \quad \cos\theta = 4/5.$$

$$\ell_2 \sin\theta = v + d\cos\theta, \quad \ell_2\cos\theta = u - d\sin\theta.$$

$$d = \frac{5}{3}u - \frac{4}{3}\ell_2, \quad v = \frac{3}{5}\ell_2 - \frac{4}{3}\left[u - \frac{4}{3}\ell_2\right].$$

This can be simplified to get

$$\begin{aligned} v &= \frac{4}{3}\ell_1 - \frac{5}{3}\ell_2, \\ &= \frac{L}{EA}\left[21P + \frac{16}{3}Q\right]. \end{aligned} \tag{3.37}$$

The flexibility matrix for this case can be seen in the relation

$$\begin{Bmatrix} u \\ v \end{Bmatrix} = \frac{L}{EA}\begin{bmatrix} 21 & \frac{16}{9} \\ \frac{16}{9} & 4 \end{bmatrix}\begin{Bmatrix} P \\ Q \end{Bmatrix}. \tag{3.38}$$

The symmetry of the matrix is obvious. When a large truss has to be analyzed, the displacements are cumulative and geometrically satisfying the compatibility of displacements will be tedious.

3.3.2 Virtual Force Method

In this method, we apply virtual forces δP and δQ at the joint C. These two forces are independent quantities, that is, we may apply them one at a time. To maintain equilibrium, the internal forces

$$\delta F_1 = \frac{4}{3}\delta P + \delta Q, \quad \delta F_2 = -\frac{5}{3}\delta P$$

are needed. From the complementary energy

$$U^* = \sum_{i=1,2} \frac{F_i^2 L_i}{2E_i A_i},$$

where, in our case, $E_i = E$ and $A_i = A$, and the virtual change in complementary energy is

$$\begin{aligned}
\delta U^* &= \frac{L}{EA}[4F_1\delta F_1 + 5F_2\delta F_2], \\
&= \frac{L}{EA}\left[4\left(\frac{4}{3}P + Q\right)\left(\frac{4}{3}\delta P + \delta Q\right) + 5\left(-\frac{5}{3}P\right)\left(-\frac{5}{3}\delta P\right)\right], \\
&= \frac{L}{EA}\left[\left(21P + \frac{16}{3}Q\right)\delta P + \left(\frac{16}{3}P + 4Q\right)\delta Q\right].
\end{aligned} \tag{3.39}$$

The virtual complementary work done by the forces is

$$\delta W^* = u\delta P + v\delta Q. \tag{3.40}$$

Equating the change in complementary energy with the work done and separating the coefficients of δP and δQ, we get

$$\begin{Bmatrix} u \\ v \end{Bmatrix} = \frac{L}{3EA}\begin{bmatrix} 63 & 16 \\ 16 & 12 \end{bmatrix}\begin{Bmatrix} P \\ Q \end{Bmatrix}. \tag{3.41}$$

We note that this approach is entirely scalar and no geometric calculation is needed.

3.3.3 Castigliano's First Theorem

Consider $\delta Q = 0$ in the preceding calculation. Then,

$$u\delta P = \frac{\partial U^*}{\partial P}\delta P, \quad u = \frac{\partial U^*}{\partial P}. \tag{3.42}$$

When $\delta P = 0$, we find

$$v = \frac{\partial U^*}{\partial Q}. \tag{3.43}$$

The procedure consists of writing U^* in terms of applied loads P and Q by eliminating the member forces F_1 and F_2 (using the equilibrium equations) in the energy expression. Differentiating with respect to P and Q gives u and v, respectively.

3.3.4 Unit Load Method

In order to facilitate a systematic way of arranging the calculations, we may write

$$u = \sum_{i=1,2} \frac{\partial U^*}{\partial F_i} f_i,$$

where

$$f_i = \frac{\partial F_i}{\partial P}.$$

The quantities f_i are the member forces due to a unit load applied at C in the direction of P. We may do the calculations as shown in the Table 3.1.

Table 3.1. Unit load calculation in a tabular form

i	$\frac{\partial U^*}{\partial F_i} = \frac{F_i L_i}{E_i A_i}$	f_i	$\frac{\partial U^*}{\partial F_i} f_i$
1	$\frac{4L}{EA}\left[\frac{4}{3}P + Q\right]$	$\frac{4}{3}$	$\frac{L}{EA}\left[\frac{64}{3}P + \frac{16}{3}Q\right]$
2	$\frac{5L}{EA}\frac{5}{3}P$	$\frac{5}{3}$	$\frac{L}{EA}\frac{125}{9}P$
	$u = \frac{L}{EA}[21P + \frac{16}{3}Q]$		

3.3.5 Virtual Displacement Method

As shown in Fig. 3.6(d), we assume displacements are given at C as u and v. As shown in Fig. 3.7, the effect of u on member 1 is an extension of u. Member

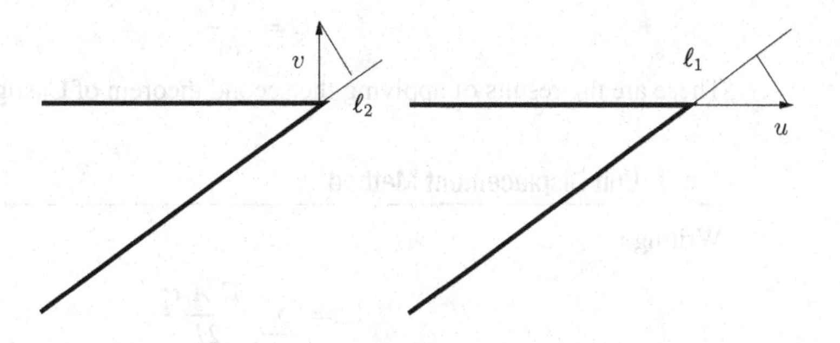

Figure 3.7 Resolving displacements to find member extensions.

2 has to rotate and elongate by $u\cos\theta$. Due to v, member 1 purely rotates with no extension and member 2 rotates and extends by $v\sin\theta$. It is to be noted that under small displacements the rotation causes the node to move perpendicular to the original direction of the structural member. Thus, the total extensions are:

$$\ell_1 = u, \quad \ell_2 = 4u/5 + 3v/5.$$

The change in strain energy is obtained as

$$\begin{aligned}\delta U &= \frac{EA}{L}\left[\frac{\ell_1\delta\ell_1}{4}+\frac{\ell_2\delta\ell_2}{5}\right],\\ &= \frac{EA}{L}\left[\frac{1}{4}u\delta u+\frac{1}{5}(4u/5+3v/5)(4\delta u/5+3\delta v/5)\right],\\ &= \frac{EA}{L}\left[\left(\left(\frac{1}{4}+\frac{16}{125}\right)u+\frac{12}{125}v\right)\delta u+\left(\frac{9}{125}v+\frac{12}{125}u\right)\delta u\right],\\ &= \frac{EA}{500L}[(189u+48v)\delta u+(48u+36v)\delta v]. \end{aligned} \tag{3.44}$$

The virtual work done by the forces is

$$\delta W = P\delta u + Q\delta v. \tag{3.45}$$

Equating the coefficients of δu and δv, in the equation $\delta W = \delta U$, we find

$$\begin{Bmatrix} P \\ Q \end{Bmatrix} = \frac{EA}{500L}\begin{bmatrix} 189 & 48 \\ 48 & 36 \end{bmatrix}\begin{Bmatrix} u \\ v \end{Bmatrix}. \tag{3.46}$$

The stiffness matrix on the right is the inverse of the flexibility we found before. We may verify this by inverting one of these matrices.

3.3.6 Castigliano's Second Theorem

In the virtual displacement method, if we assume that δu is given with $\delta v = 0$, first, we see

$$P = \frac{\partial U}{\partial u},$$

and if we let $\delta u = 0$ and $\delta v \neq 0$,

$$Q = \frac{\partial U}{\partial v}.$$

These are the results of applying the second theorem of Castigliano.

3.3.7 Unit Displacement Method

Writing

$$U = \sum_{i=1,2} \frac{E_i A_i \ell_i^2}{2L_i},$$

and using the second theorem of Castigliano,

$$P = \sum_{i=1,2} \frac{\partial U}{\partial \ell_i} g_i,$$

where

$$g_i = \frac{\partial \ell_i}{\partial u}.$$

The quantity g_i can be seen as the extension of the member i due to a unit displacement in the direction of u. We may compute the force P using Table 3.2.

Table 3.2. Unit displacement calculation in a tabular form

i	$\frac{\partial U}{\partial \ell_i} = \frac{E_i A_i \ell_i}{L_i}$	g_i	$\frac{\partial U}{\partial \ell_i} g_i$
1	$\frac{EA}{4L} u$	1	$\frac{EA}{L} \frac{1}{4} u$
2	$\frac{EA}{5L}\left[\frac{4}{5}u + \frac{3}{5}v\right]$	$\frac{4}{5}$	$\frac{EA}{L}\left[\frac{16}{125}u + \frac{12}{125}v\right]$
	$P = \frac{EA}{500L}[189u + 48v]$		

3.3.8 Minimum Total Complementary Potential Energy Theorem

The prescribed displacements u and v may be expressed in terms of a complementary potential energy, as

$$u = -\frac{\partial V^*}{\partial P}, \quad v = -\frac{\partial V^*}{\partial Q},$$

where V^* is called the complementary potential energy of the displacements. Then Castigliano's first theorem states

$$\frac{\partial U^*}{\partial P} = -\frac{\partial V^*}{\partial P}, \quad \frac{\partial U^*}{\partial Q} = -\frac{\partial V^*}{\partial Q}.$$

By defining a total complementary potential energy function

$$\Pi^* = U^* + V^*,$$

Castigliano's theorem can be expressed as

$$\frac{\partial \Pi^*}{\partial P} = 0, \quad \frac{\partial \Pi^*}{\partial Q} = 0,$$

which shows Π^* is a minimum when compatibility is satisfied. As U^* is a positive definite quadratic function, this is indeed a minimum condition and not a maximum condition.

3.3.9 Minimum Total Potential Energy Theorem

The prescribed forces P and Q are expressed as

$$P = -\frac{\partial V}{\partial u}, \quad Q = -\frac{\partial V}{\partial v},$$

where V is the potential energy of the forces. Now, Castigliano's second theorem becomes

$$\frac{\partial \Pi}{\partial u} = 0, \quad \frac{\partial \Pi}{\partial v} = 0,$$

where Π is the total potential energy given by

$$\Pi = U + V.$$

When the equilibrium equations are satisfied the total potential energy is a minimum.

3.3.10 Dummy Load and Dummy Displacements

Assuming in our two-bar truss problem only the load P is applied, to find the vertical deflection v we are left with no variable Q to differentiate U^* with. In this situation we apply a *dummy* load Q in the direction of v and after differentiation we set the dummy load to zero. We can follow the same procedure if in the case of displacement u is given and vertical displacement is given as $v = 0$. We provide a dummy displacement v to construct U and after differentiation, it is set to zero.

3.4 Statically Indeterminate Problems

In redundant structures subject to nodal forces, the number of equilibrium equations are not sufficient in number to find all the member forces. We call these problems *statically indeterminate*. As an example, consider the truss shown in

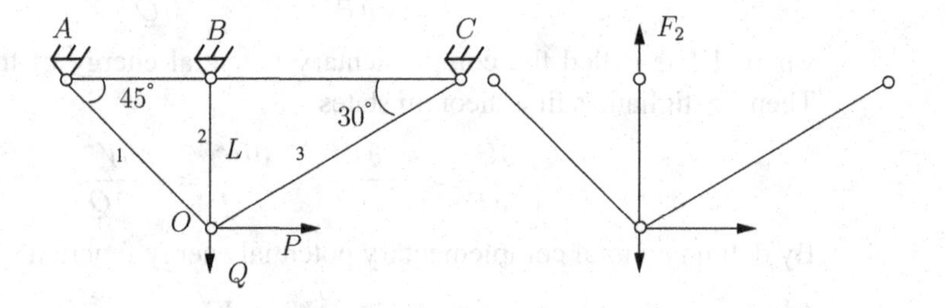

Figure 3.8 A statically indeterminate truss.

Fig. 3.8. We assume EA is common for all three members. At the node O the equilibrium equations are

$$F_1 \cos 45 - F_3 \cos 30 = P$$

$$F_1 \sin 45 + F_2 + F_3 \sin 30 = Q.$$

We have three unknowns and two equations. One way to proceed is to solve for F_1 and F_3 in terms of P, Q, and F_2. This leads to

$$F_1 = \frac{\sqrt{2}}{1+\sqrt{3}}\left[P + \sqrt{3}(Q - F_2)\right],$$

$$F_3 = \frac{2}{1+\sqrt{3}}[Q - F_2 - P].$$

The complementary energy is obtained as

$$U^* = \frac{L}{EA}\left[\sqrt{2}F_1^2 + F_2^2 + 2F_3^2\right],$$

$$= \frac{L}{EA}\left[\frac{2\sqrt{2}}{(1+\sqrt{3})^2}(P + \sqrt{3}Q - \sqrt{3}F_2)^2 + F_2^2 + \frac{8}{(1+\sqrt{3})^2}(P - Q + F_2)^2\right].$$

From Fig. 3.8 we observe that the force F_2 does not do any work. Then

$$\frac{\partial U^*}{\partial F_2} = 0.$$

This gives the needed additional equation to solve for F_2 in terms of the applied loads.

If we use the displacement method we can find the extensions of all the members in terms of the nodal displacements at O. The problem of static indeterminacy does not arise in the displacement method. In fact, the degree of static indeterminacy is defined as the difference between the number of unknown member forces and the number of equilibrium equations. When this degree is large, it is always advantageous to use the displacement method.

3.4.1 Example: Statically Indeterminate Torsion

Consider a shaft with torsional rigidity GJ clamped between two rigid walls (Fig. 3.9). A torque T is applied at a distance a from the left wall. The free-body

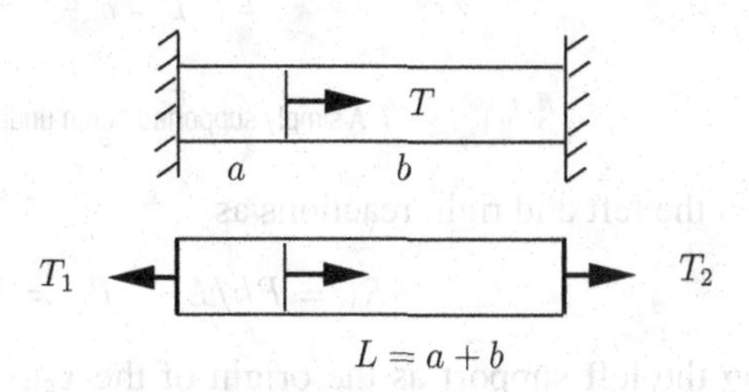

Figure 3.9 A statically indeterminate shaft.

diagram shows that there are two unknown torques T_1 and T_2 reacting from the walls. We have one equilibrium equation:

$$T + T_2 - T_1 = 0.$$

The complementary energy is

$$U^* = \frac{1}{2GJ}\left[aT_1^2 + bT_2^2\right].$$

Eliminating T_1 using the equilibrium equation:

$$U^* = \frac{1}{2GJ}\left[a(T + T_2)^2 + bT_2^2\right].$$

As T_2 does not do any work, $\partial U^*/\partial T_2 = 0$. This gives

$$LT_2 = -aT, \quad U^* = \frac{1}{2GJ}\left[a(T - aT/L)^2 + ba^2T^2/(L^2)\right] = \frac{abT^2}{2GJL}.$$

The rotation of the shaft under the torque is obtained as

$$\theta = \frac{\partial U^*}{\partial T} = \frac{abT}{GJL}.$$

For this problem, if we use the displacement method,

$$U = \frac{GJ}{2}\left[\frac{\theta^2}{a} + \frac{\theta^2}{b}\right].$$

The torque is obtained as

$$T = \frac{\partial U}{\partial \theta} = \frac{GJL\theta}{ab}.$$

3.4.2 Example: Beam under a Concentrated Load

Consider a beam of bending stiffness EI under a load P. We are interested in finding the deflection under the load. By taking moments about supports we

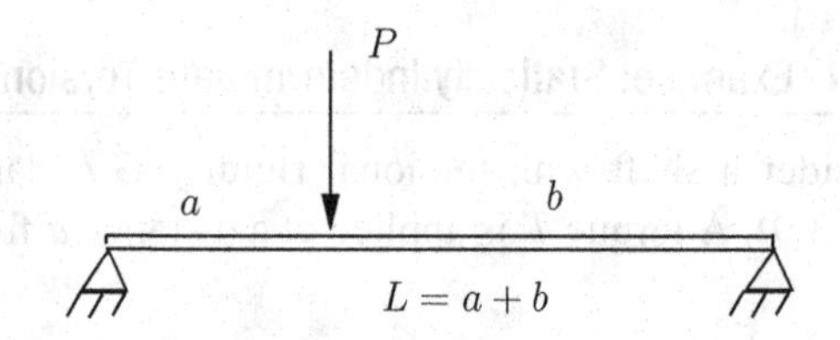

Figure 3.10 A simply supported beam under a concentrated load.

obtain the left and right reactions as

$$R_1 = Pb/L, \quad R_2 = Pa/L.$$

Using the left support as the origin of the x-axis, the bending moment in $0 < x < a$ is

$$M_1 = R_1 x = Pbx/L.$$

The bending moment in $a < x < L$ is given by

$$M_2 = R_2(L - x) = Pa(L - x)/L.$$

The complementary energy stored is found as

$$U^* = \frac{1}{2EI}\left[\int_0^a M_1^2 dx + \int_a^L M_2^2 dx\right],$$

$$= \frac{1}{2EIL^2}\left[\frac{b^2a^3}{3} + \frac{a^2b^3}{3}\right]P^2,$$

$$= \frac{P^2a^2b^2}{6EIL}.$$

The deflection under the load v is found as

$$v = \frac{\partial U^*}{\partial P} = \frac{Pa^2b^2}{3EIL}.$$

In this example we used a single coordinate x from left to right. Using a second coordinate from right to left to express M_2 is possible.

If we want to use the displacement method we have a family of choices for the deflection $v(x)$ subjected to the admissibility conditions: they are continuous with continuous derivatives, they are zero at the support points, and they have a prescribed value (say, v_0) under the load. Out of these infinite choices, the one that makes the total potential energy Π a minimum is the true solution. If we pick any arbitrary shape, meeting the stated conditions, Π will be higher than the actual minimum value, and we get a higher value for P, implying a stiffer beam. For example, let us try a parabolic shape,

$$v(x) = Vx(L-x), \quad v(a) = v_0 = Vab, \quad V = v_0/(ab).$$

Then

$$v(x) = \frac{v_0}{ab}x(L-x)$$

meets all the admissibility conditions. The strain energy U can be obtained as

$$U = \frac{EI}{2}\int_0^L (v'')^2 dx = \frac{2EIL}{a^2b^2}v_0^2.$$

Using $P = \partial U/\partial v_0$,

$$P_{approx} = \frac{4EIL}{a^2b^2}v_0.$$

Here, we get a factor of 4 instead of the exact value 3. We may improve our result by systematically using improved displacements. For example,

$$v(x) = \frac{v_0}{ab}x(L-x)[1 + A_1(x-a) + A_2(x-a)^2 + \cdots].$$

This has additional degrees of freedom and U has to be minimized with respect to A_1, A_2, and so on. This approach is called the Rayleigh-Ritz method. The modern approach to solving elasticity problems is the finite element method. For our example, the beam will be considered as a collection of smaller beams assembled together with common deflections and slopes between the joined ends of two adjacent beam elements. These deflections and slopes form additional degrees of freedom in minimizing the potential energy. There are a number of software packages available to accomplish this task.

3.4.3 Example: Axial Extension Calculation of a Variable Cross-section Bar by the Finite Element Method

For the bar shown in Fig. 3.11, the cross-section varies in the form

$$A = A_0 e^{-x/L},$$

where L is the length. The Young's modulus E is constant. The co-ordinate x is taken from left to right. At $x = 0$ the bar is clamped to a wall and there is no displacement. The other end, $x = L$, is given an elongation $\bar{u}$. First, let us find an

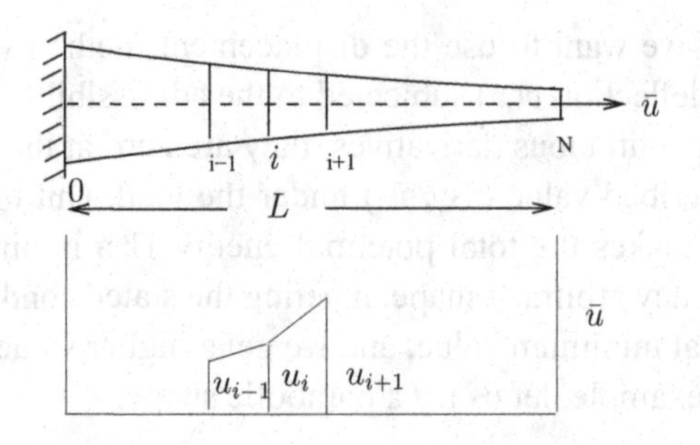

Figure 3.11 A bar sectioned into finite elements.

exact solution for this problem. The end displacement $\bar{u}$ is caused by an applied force P. Then the stress and strain at any section x is given by

$$\sigma = P/A, \quad \epsilon = P/(EA).$$

Using $\epsilon = du/dx$ and $u((0) = 0$, we find

$$u(x) = \frac{P}{EA_0}\int_0^x e^{x/L}dx = \frac{PL}{EA_0}[e^{x/L} - 1].$$

From this,

$$\bar{u} = \frac{PL}{EA_0}(e - 1).$$

The exact value of the load is

$$P_{ex} = \frac{EA_0\bar{u}}{L(e-1)},$$

and the exact displacement distribution is

$$u_{ex}(x) = \bar{u}\frac{e^{x/L} - 1}{e - 1}.$$

Now, if we use the displacement method, we have to choose a displacement that satisfies the admissibility conditions: $u(0) = 0$, $u(L) = \bar{u}$ and continuity. Following the finite element method, we divide the length L into N elements. Let

$$h = L/N.$$

Then, the ith element is between $x = (i-1)h$ and $x = ih$. We assume the displacement is distributed linearly inside this element with the value of $u = u_{i-1}$ at the left end and $u = u_i$ at the right end. In this element the slope of the displacement $u' = (u_i - u_{i-1})/h$. The displacement distribution in two adjoining elements is shown in Fig. 3.11. We have the admissibility conditions, $u_0 = 0$ and $u_N = \bar{u}$. We will have $(N-1)$ unknowns, $u_1, u_2, \ldots, u_{N-1}$ to find. The strain energy can be expressed as

$$
\begin{aligned}
U &= \frac{1}{2}\int_0^L EA(u')^2 dx, \\
&= \frac{EA_0}{2}\sum_{i=1}^{N}\int_{(i-1)h}^{ih} e^{-x/L}\frac{(u_i - u_{i-1})^2}{h^2}dx, \\
&= \frac{EA_0 L}{2h^2}\sum_{i=1}^{N}(u_i - u_{i-1})^2[e^{-(i-1)h/L} - e^{-ih/l}], \\
&= \frac{EA_0 L}{2h^2}(e^{h/L} - 1)\sum_{i=1}^{N} e^{-ih/L}(u_i - u_{i-1})^2, \\
&= \frac{EA_0 N^2}{2L}(e^{1/N} - 1)\sum_{i=1}^{N} e^{-i/N}(u_i - u_{i-1})^2.
\end{aligned}
$$

This can be written as

$$
U = \frac{EA_0}{2L}\sum_{i=1}^{N} I_i(u_i - u_{i-1})^2,
$$

where

$$
I_i = N^2(e^{1/N} - 1)e^{-i/N}.
$$

If we write this explicitly, we have

$$
U = \frac{EA_0}{2L}\left[I_1 u_1^2 + I_2(u_2 - u_1)^2 + I_3(u_3 - u_2)^2 + \cdots + I_N(\bar{u} - u_{N-1})^2\right].
$$

Using this in the total potential energy $\Pi = U - \bar{u}P$ and minimizing Π with $u_i, i = 1, 2, \ldots, N-1$, we get

$$
(I_i + I_{i+1})u_i - I_i u_{i-1} - I_{i+1}u_{i+1} = 0, \quad i = 1, 2, \ldots N-1. \tag{3.47}
$$

Differentiating with $\bar{u}$, we get

$$
P = \frac{EA_0}{L} I_N(\bar{u} - u_{N-1}). \tag{3.48}
$$

The crudest approximation we can make is:

$$
N = 1.
$$

Then

$$
I_N = \frac{e-1}{e}, \quad P = \frac{EA_0}{L}\frac{e-1}{e}\bar{u}.
$$

We find

$$
\frac{P}{P_{ex}} = \frac{(e-1)^2}{e} = 1.0862.
$$

As mentioned earlier, displacement approximation leads to a stiffer structure.

3.4.4 Example: Frame Deflection

When beam elements are joined with rigid corners as shown in Fig. 3.12, we have what are called frames. Here, let us try to compute the horizontal displacementat

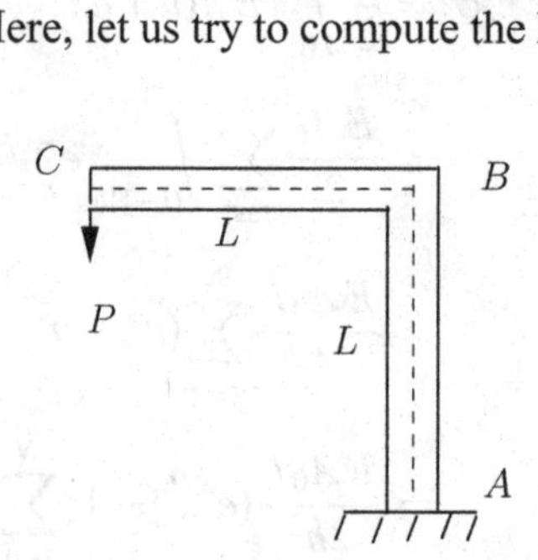

Figure 3.12 A frame with a vertical load.

B due to the applied vertical load P at C. We assume EI is a constant along the frame. In this example, we neglect the deflections due to axial forces. Compared to bending deflections, these are usually negligible.

As there is no applied load at B, we apply a dummy load Q (pointing to the left). The bending moment between C and B is found using a coordinate x to the right from C as

$$M_1 = -Px.$$

For the segment AB, we use a new x starting from B going downward. Then

$$M_2 = -(PL + Qx).$$

The complementary energy is

$$U^* = \frac{1}{2EI}\int_0^L \left(M_1^2 + M_2^2\right)^2 dx,$$

$$= \frac{L^3}{2EI}\left[\frac{P^2}{3} + P^2 + PQ + \frac{Q^2}{3}\right],$$

$$u = \left.\frac{\partial U^*}{\partial Q}\right|_{Q=0} = \frac{PL^3}{2EI}.$$

3.5 Example: Bulkhead Subjected to Moments

A bulkhead can be treated as a beam if its cross-sectional dimensions are small compared to its radius. Fig. 3.13(a) shows the moment loadings, M_0, at diametrically opposite points. This may be viewed as an approximation for the moments transmitted by the wings on to a fuselage bulkhead. We assume the bulkhead has a bending stiffness EI and radius R. Our objective is to find the angle of rotation, β, experienced by each moment.

We begin by cutting the bulkhead into two parts along a diameter as shown in Fig. 3.13(b) and (c). The applied moment M_0 is split into M_1 applied on

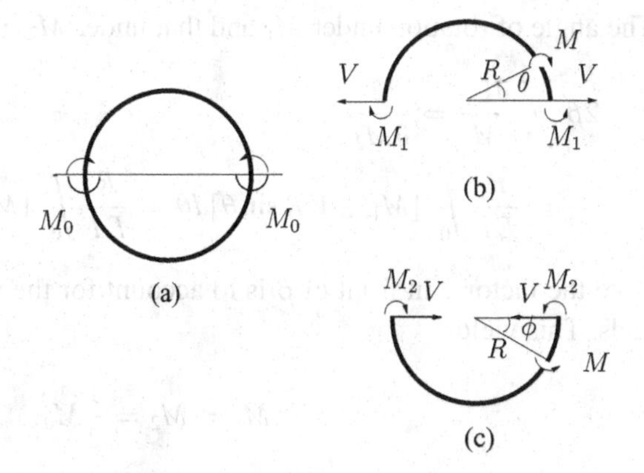

Figure 3.13 Bulkhead under symmetrically applied moments.

Fig. 3.13(b) and M_2 applied to Fig. 3.13(c). We also apply equal and opposite shear forces V at the cuts. There are no tangential forces due to the symmetry. We have to satisfy

$$M_1 + M_2 = M_0. \tag{3.49}$$

If the semicircle is cut at an arbitrary angle θ, at the cut we have to insert a moment M, a shear force, and a tangential force. If we take the moment about a point at θ on the bulkhead, the shear force and the tangential force do not contribute to the bending moment. For the two semicircles the bending moments are:

$$M(\theta) = M_1 + VR\sin\theta, \quad M(\phi) = -M_2 - VR\sin\phi, \tag{3.50}$$

where ϕ is the angle for describing the lower half.

The total complementary energy stored in the bulkhead is

$$\begin{aligned} U &= \frac{1}{2EI}\left\{\int_0^\pi M^2(\theta)Rd\theta + \int_0^\pi M^2(\phi)Rd\phi\right\}, \\ &= \frac{R}{2EI}\left\{\int_0^\pi [M_1 + VR\sin\theta]^2 d\theta + \int_0^\pi [M_2 + VR\sin\phi]^2 d\phi\right\}. \end{aligned} \tag{3.51}$$

The work done by the internal force V should be zero. That is,

$$\frac{\partial U}{\partial V} = 0. \tag{3.52}$$

Then

$$\int_0^\pi [M_1 + VR\sin\theta]\sin\theta d\theta + \int_0^\pi [M_2 + VR\sin\phi]\sin\phi d\phi = 0. \tag{3.53}$$

This expression, after integration, gives

$$V = -\frac{2M_0}{\pi R}. \tag{3.54}$$

The angle of rotation under M_1 and that under M_2 must be the same.

$$2\beta = \frac{\partial U}{\partial M_1} = \frac{\partial U}{\partial M_2}$$
$$= \frac{R}{EI}\int_0^{\pi}[M_1 + VR\sin\theta]d\theta = \frac{R}{EI}\int_0^{\pi}[M_2 + VR\sin\phi]d\phi. \quad (3.55)$$

Note the factor 2 in front of β is to account for the work done by M_1 at the two ends. This yields

$$M_1 = M_2 = \frac{1}{2}M_0. \quad (3.56)$$

and

$$\beta = \frac{\pi M_0 R}{4EI}\left(1 - \frac{8}{\pi^2}\right). \quad (3.57)$$

Using M_1, M_2, and V, the bending moments and the bending stresses could be found at any point in the bulkhead.

3.6 Example: Shear Correction Factor

Consider the cantilever beam of length L shown in Fig. 3.14. It has a width b and height $2c$. The second moment of area is

$$I = 2bc^3/3.$$

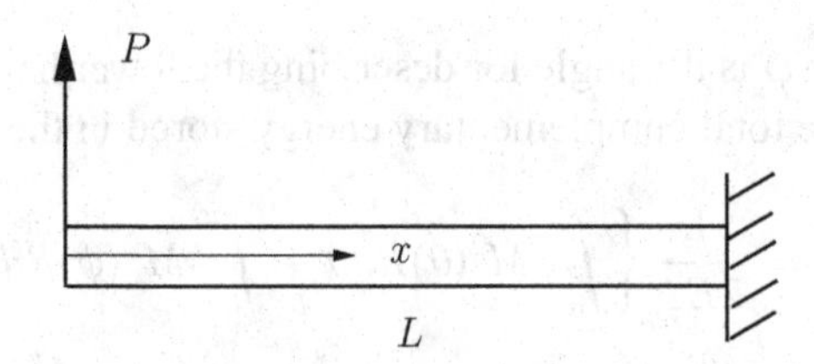

Figure 3.14 A cantilever beam with a tip load.

The bending moment and normal stress at a section x are

$$M = -Px, \quad \sigma_{xx} = -Pxy/I.$$

From the equilibrium equation (neglecting body forces),

$$\sigma_{xx,x} + \sigma_{xy,y} = 0, \quad \sigma_{xy,y} = Py/I.$$

Integrating this we get

$$\sigma_{xy} = \frac{P}{2I}(y^2 - c^2).$$

The complementary energy due to bending and shear can be written as

$$U^* = \int_0^L \frac{M^2}{2EI} dx + \int_0^L b \int_{-c}^{c} \frac{\sigma_{xy}^2}{2G} dy dx,$$

$$= \frac{P^2}{2EI} \int_0^L x^2 dx + 2bL \frac{P^2}{8GI^2} \int_0^c (y^2 - c^2)^2 dy,$$

$$= \frac{P^2 L^3}{6EI} + bL \frac{P^2}{4GI^2} c^5 \left[\frac{1}{5} - \frac{2}{3} + 1 \right],$$

$$= \frac{P^2 L^3}{6EI} \left[1 + \frac{6Ebc^5}{4GIL^2} \frac{8}{15} \right],$$

$$= \frac{P^2 L^3}{6EI} \left[1 + \frac{6}{5} \frac{Ec^2}{GL^2} \right].$$

The deflection v under the load is given by

$$v = \frac{PL^3}{3EI} \left[1 + \frac{6}{5} \frac{Ec^2}{GL^2} \right].$$

Here, the factor outside the brackets represents the contribution from the bending moment and the shear deformation effect is represented by the term containing the (6/5) factor, which is of the order of c^2/L^2.

3.7 Reciprocity Theorem

Consider a linearly elastic structure subjected to two loads, P_1 and P_2. Let u_1 and u_2 denote the deflection under the two loads, respectively. Then, we have load-displacement relations

$$u_1 = c_{11} P_1 + c_{12} P_2,$$

$$u_2 = c_{21} P_1 + c_{22} P_2.$$

From Castigliano's theorem,

$$u_1 = \frac{\partial U^*}{\partial P_1}, \quad u_2 = \frac{\partial U^*}{\partial P_2}.$$

Using the equality of the mixed derivatives of U^*, we get

$$\frac{\partial u_1}{\partial P_2} = \frac{\partial u_2}{\partial P_1}.$$

This results in the reciprocity relation $c_{12} = c_{21}$. In other words, the incremental deflection at point 1 due to a load P at point 2 is the same as the incremental deflection at 2 due to the same load at 1.

A similar relation between the stiffness coefficients, $k_{12} = k_{21}$, can be obtained from the strain energy U.

FURTHER READING

Argyris, J. H. and Kelsey, S., *Energy Theorems and Structural Analysis*, Butterworth (1960).

Curtis, H. D., *Fundamentals of Aircraft Structural Analysis*, Irwin (1997).

Megson, T. H. G., *Aircraft Structures for Engineering Students*, Butterworth-Heinemann (2007).

Timoshenko, S. P. and Gere, J. M., *Mechanics of Materials*, PWS Publishers (1984).

EXERCISES

3.1 Four rigid bars are connected by pin joints as shown in Fig. 3.15. Using the principle of virtual work, find the value of P needed for equilibrium. To begin the solution, give a (small) virtual displacement u to the joint where P is applied. This will deform the square into a parallelogram. The virtual displacement of the vertical load can be obtained from geometry.

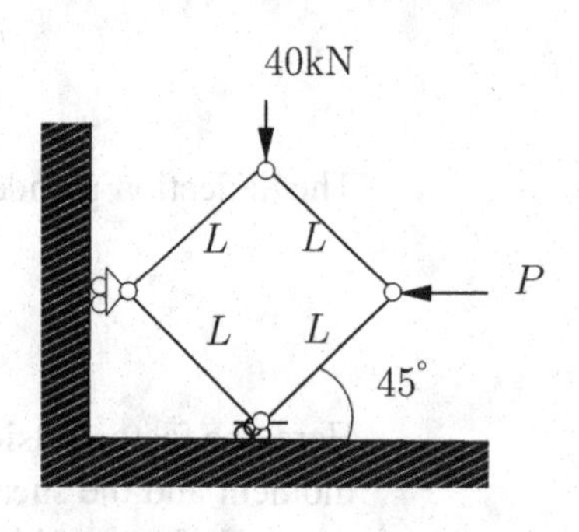

Figure 3.15 Pin-connected rigid bars.

3.2 An X-shaped support mechanism is shown in Fig. 3.16. Obtain the angle θ when it is in equilibrium with the two loads shown. The solution should begin with a virtual change in angle $\delta\theta$ and the ensuing virtual displacements of the weight and the applied load. All bars are rigid.

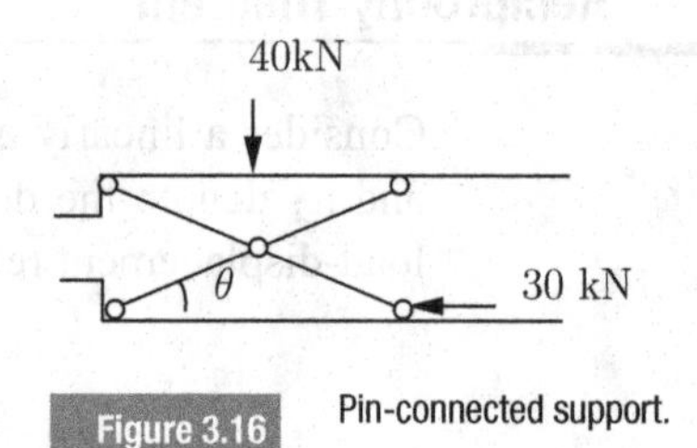

Figure 3.16 Pin-connected support.

3.3 A three-bar truss shown in Fig. 3.17 has a horizontal bar of cross-sectional area $2A$ and the inclined bars have A. All of them have Young's modulus E. Obtain the displacements u along the direction of the applied load P and v along the load Q, using Castigliano's first theorem.

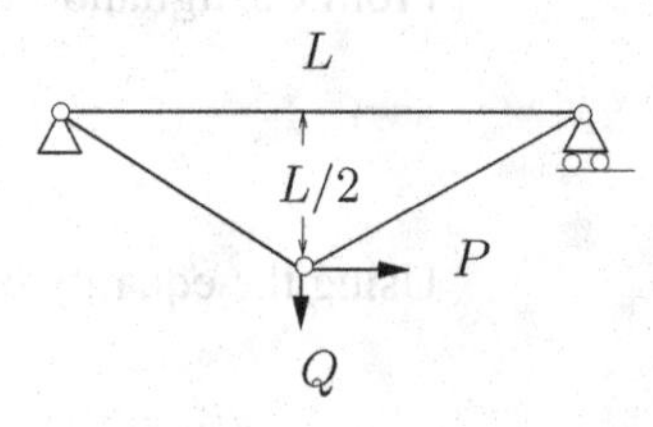

Figure 3.17 A three-bar truss.

3.4 For the truss shown in Fig. 3.17, compute the displacements u and v along the directions of P and Q, respectively, using Castigliano's second theorem.

3.5 Under an applied load of Q obtain the vertical and horizontal deflection at the point of its application. The two bars in Fig. 3.18, are identical with cross-sectional areas A and the Young's modulus E.

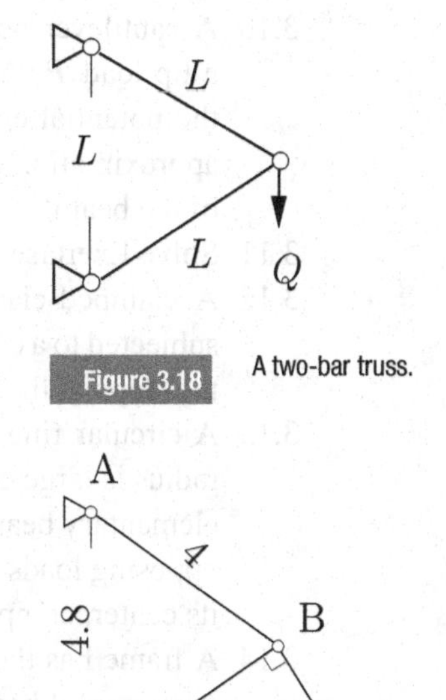

Figure 3.18 A two-bar truss.

3.6 The truss shown in Fig. 3.19 is made of elastic bars with $E = 200$ GPa and cross-sectional areas $A = 500\,\text{mm}^2$. The dimensions of the truss are in meters. The applied load $Q = 2.4\,\text{kN}$. Obtain the vertical and horizontal displacements at C.

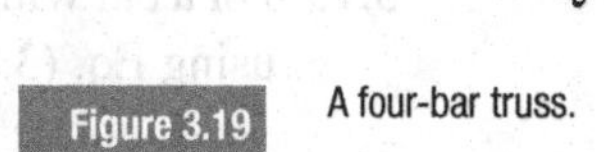

Figure 3.19 A four-bar truss.

3.7 The square truss in Fig. 3.20 has overlapping cross braces. The diagonals are not connected at the center. Each member has a cross-sectional area of $500\,\text{mm}^2$ and Young's modulus 200 GPa. The dimensions shown are in meters. The applied load P at B is 100 kN. Obtain the deflection along the direction of the load using the displacement method.

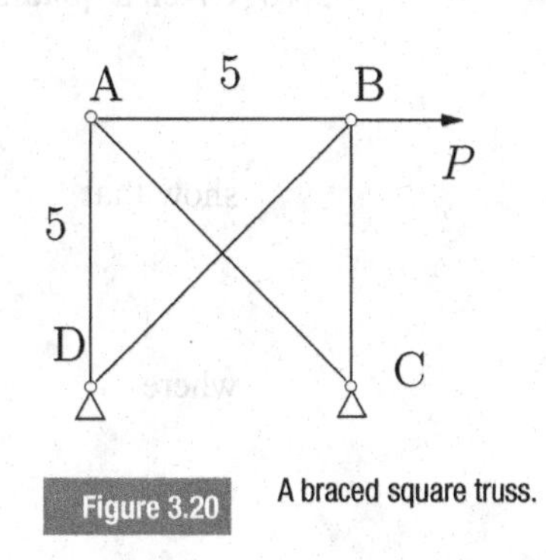

Figure 3.20 A braced square truss.

3.8 For the truss in Exercise 3.7, obtain the deflection under the load using the force method.

3.9 A bicycle wheel is idealized as a rigid circle resting on the ground, with 24 spokes of length 40 cm distributed evenly around. This is shown in Fig. 3.21. If the cross-sectional areas of the spokes are $0.15\,\text{mm}^2$ and $E = 200$ GPa, compute the vertical deflection under the hub when a load 2 kN is applied vertically at the hub.

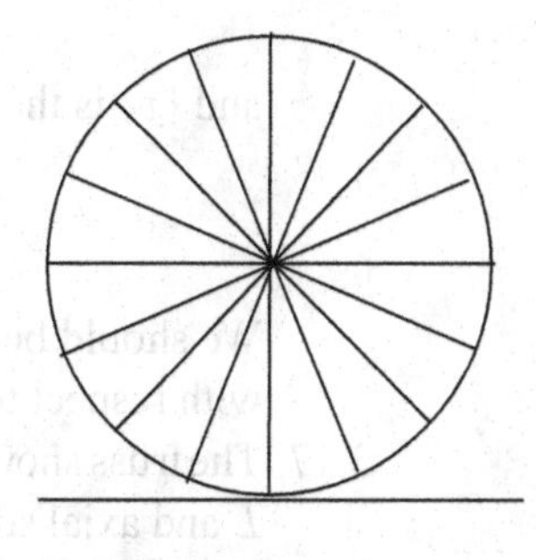

Figure 3.21 A bicycle wheel model.

3.10 A cantilever beam of length L and bending stiffness EI is subjected to a tip load P. Assuming a displacement distribution Ax^2 and minimizing the potential energy, obtain the tip deflection for the beam. Improve this approximation using $Ax^2 + Bx^3$. Here, x is measured from the fixed end of the beam.

3.11 Solve Exercise 3.10 using the complementary energy method.

3.12 A clamped-clamped beam with length $2L$ and bending stiffness EI is subjected to a concentrated clockwise moment M_0 at its mid-span. Compute the slope at the mid-span using the complementary energy.

3.13 A circular ring of radius R has a bending stiffness of EI. Assuming the radius is large compared to the cross-sectional dimensions, we may use the elementary beam theory to deal with the ring. If the ring is subjected to two opposing loads P at diametrically opposite points, both directed towards to its center, compute the deflections under the loads.

3.14 A frame has the shape of a gallows with a vertical beam of length $2L$ and a horizontal beam of length L. If the bending stiffness is EI for the both parts, compute the vertical deflection under a load P applied vertically at the tip.

3.15 For a bar with exponentially decreasing cross-section discussed in the book, using Eq. (3.47) in Eq. (3.48), compute the load P for $N = 2, 4, 6, 8$, and 10. Tabulate P/P_{ex} to show the convergence of the method.

3.16 Given a quadratic form

$$U = \frac{1}{2}(a_{11}x_1^2 + 2a_{12}x_1x_2 + a_{22}x_2^2), \tag{3.58}$$

show that

$$U = \frac{1}{2}\{x\}^T[A]\{x\}, \tag{3.59}$$

where

$$\{x\} = \begin{Bmatrix} x_1 \\ x_2 \end{Bmatrix}, \quad [A] = \begin{bmatrix} a_{11} & a_{12} \\ a_{12} & a_{22} \end{bmatrix}. \tag{3.60}$$

If

$$y_1 = \frac{\partial U}{\partial x_1}, \quad y_2 = \frac{\partial U}{\partial x_2}, \tag{3.61}$$

and $\{y\}$ is the column matrix formed using the y-components, show that

$$\{y\} \equiv \frac{\partial U}{\partial \boldsymbol{x}} = [A]\{x\}. \tag{3.62}$$

We should be familiar with the notation for the derivative of a function U with respect to a vector $\boldsymbol{x}$.

3.17 The truss shown in Fig. 3.22, is made of equilateral triangles with side length L and axial stiffness EA. Movable nodes are numbered starting with $j = 1$ and the bars starting with $i = 1$. At a node j, assume there are applied horizontal and vertical external loads P_j and Q_j and the corresponding displacements u_j and v_j.

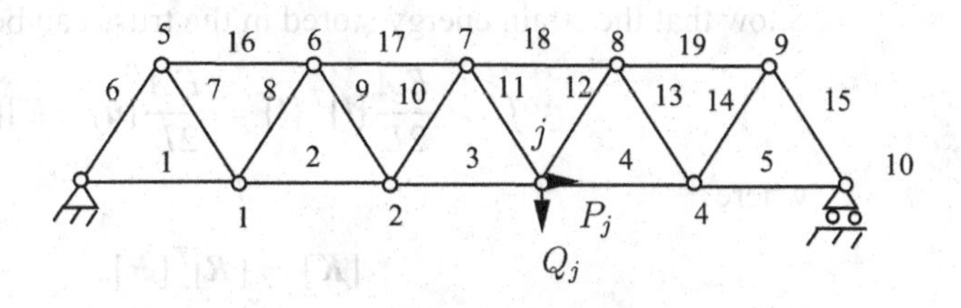

Figure 3.22 A multi-bar truss.

Write the equilibrium equations at each node relating the member forces to the external forces. Arrange them in the matrix form

$$[\boldsymbol{B}]\{\boldsymbol{F}\} = \{\boldsymbol{P}\},$$

where

$$\{\boldsymbol{F}\} = \begin{Bmatrix} F_1 \\ F_2 \\ .. \\ F_{19} \end{Bmatrix}, \quad \{\boldsymbol{P}\} = \begin{Bmatrix} P_1 \\ Q_1 \\ .. \\ P_{10} \end{Bmatrix}. \tag{3.63}$$

Show that the complementary energy stored in the truss can be written as

$$U^* = \frac{L}{2EA}\{\boldsymbol{F}\}^T\{\boldsymbol{F}\} = \frac{L}{2EA}\{\boldsymbol{P}\}^T[\boldsymbol{C}]\{\boldsymbol{P}\}, \tag{3.64}$$

where

$$[\boldsymbol{C}] = [\boldsymbol{D}]^T[\boldsymbol{D}], \quad [\boldsymbol{D}] = [\boldsymbol{B}]^{-1}. \tag{3.65}$$

Using Castigliano's first theorem, show that

$$\{\boldsymbol{u}\} = \frac{L}{EA}[\boldsymbol{C}]\{\boldsymbol{P}\}, \tag{3.66}$$

where

$$\{\boldsymbol{u}\} = \begin{Bmatrix} u_1 \\ v_1 \\ \dots \\ u_{11} \end{Bmatrix}. \tag{3.67}$$

Compute the matrix $[\boldsymbol{C}]$ and obtain u_{10} due to $Q_7 = 1$; while all other loads are zero.

3.18 For the truss in Fig. 3.22, if the nodal displacements and the member stretches are arranged in columns as

$$\{\boldsymbol{u}\} = \begin{Bmatrix} u_1 \\ v_1 \\ .. \\ u_{10} \end{Bmatrix}, \quad \{\boldsymbol{l}\} = \begin{Bmatrix} l_1 \\ l_2 \\ .. \\ l_{19} \end{Bmatrix}, \tag{3.68}$$

obtain the matrix $[\boldsymbol{R}]$, which relates the displacements to stretches in the form

$$\{\boldsymbol{l}\} = [\boldsymbol{R}]\{\boldsymbol{u}\}.$$

Show that the strain energy stored in the truss can be written as

$$U = \frac{EA}{2L}\{\boldsymbol{l}\}^T\{\boldsymbol{l}\} = \frac{EA}{2L}\{\boldsymbol{u}\}^T[\boldsymbol{K}]\{\boldsymbol{u}\}, \tag{3.69}$$

where

$$[\boldsymbol{K}] = [\boldsymbol{R}]^T[\boldsymbol{R}]. \tag{3.70}$$

Using Castigliano's second theorem, show that

$$\{\boldsymbol{P}\} = \frac{EA}{L}[\boldsymbol{K}]\{\boldsymbol{u}\}. \tag{3.71}$$

Compute the matrix $[\boldsymbol{K}]$ and verify $[\boldsymbol{K}][\boldsymbol{C}] = [\boldsymbol{I}]$, the identity matrix. Also, compute the displacement u_{10} due to the load $Q_7 = 1$, while all other loads are zero.

3.19 Two bars with stiffness EA and length L are pin-connected and lie horizontally prior to loading. A vertical load is applied at the mid-point and slowly increased to its current value P, while the bars have stretched and they appear as the dashed lines shown in Fig. 3.23. Assume the angle β is sufficiently small to use the approximations $\sin\beta = \beta$ and $\cos\beta = 1 - \beta^2/2$. Compute the strain energy stored in the bars in terms of the vertical deflection δ under the load. Obtain a relation between applied load and the deflection. Also, compute the complementary energy in terms of P. Obtain the displacement. Show that it is different from what is obtained from strain energy. This illustrates that the complementary energy method fails when there is geometric nonlinearity in the problem.

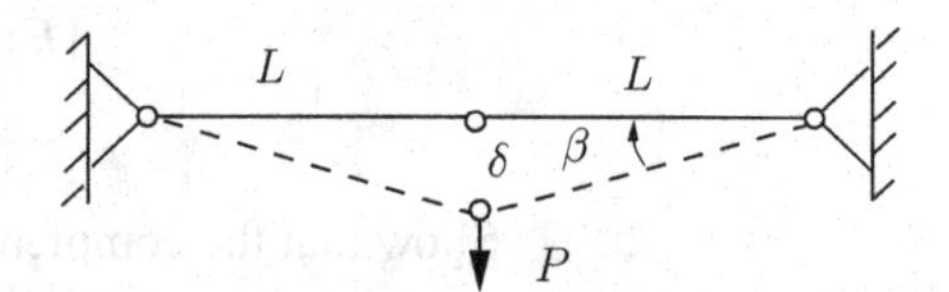

Figure 3.23 Geometrically nonlinear deformation of a two-bar truss.

3.20 A coil spring has a mean radius R and is made from a steel wire of radius r. The helix with radius R has the equation $x = R\theta$ and the pitch $p = 2\pi R$.

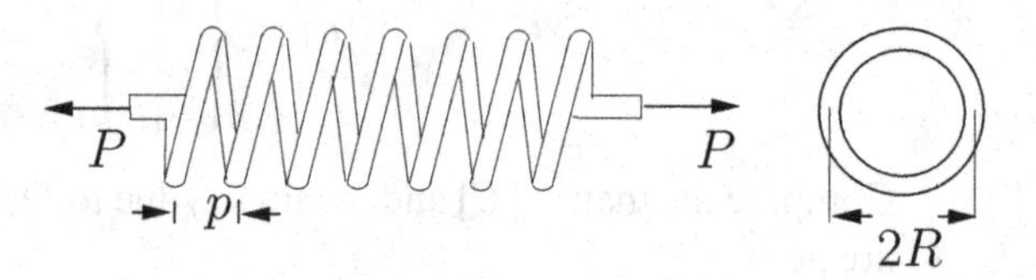

Figure 3.24 A coil spring.

For the axial load P, on the wire cross-section there is a moment $M = PR$ perpendicular to the axis of the spring. If we resolve this moment parallel to the cross-section we get a bending moment on the wire and the remaining component perpendicular to it gives a torque. Using the complementary energy stored in n turns of the spring, obtain the spring constant k.

4 Torsion

In this chapter, we begin with a solid circular shaft under torsion. Then we specialize it to a thin-walled tube. Arbitrary cross-section single cell and multi-cell tubes will be considered after it. We conclude with open-cell cross-sections and warping deformation of tubes.

4.1 Solid Circular Shaft

Let us recall the torsion of a solid circular shaft of radius c and length L. We use the axis of the shaft as the z-axis. The basic assumption concerning deformation is that cross-sections of the shaft perpendicular to its axis rotate as a rigid body under the applied torque T. If the relative rotation of the right end with respect

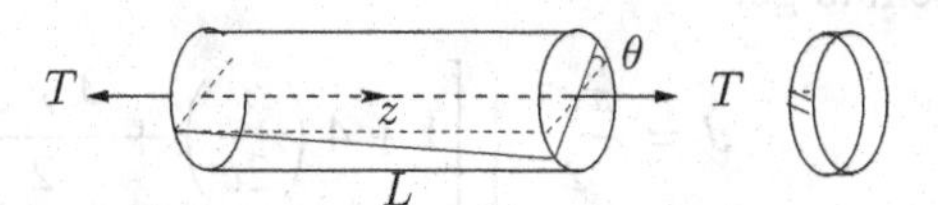

Figure 4.1 Circular shaft under torsion.

to the left end is θ, a dashed line on the surface moves to a solid helical curve, as shown in Fig. 4.1. The surface elements undergo a shear strain of

$$\gamma_{max} = c\theta/L.$$

This distortion is shown on a slice below the shaft in the figure. Interpolating this into the interior using the rigid body assumption, the shear strain at a radius r is

$$\gamma = r\theta/L.$$

The quantity

$$\alpha \equiv \theta/L$$

is called the rate of twist. Thus

$$\gamma = \alpha r, \quad \tau = G\alpha r,$$

where τ is the shear stress. A relation between the applied torque and the rate of twist is obtained from

$$T = \int_A r\tau dA = 2\pi \int_0^c G\alpha r^3 dr = GJ\alpha,$$

where

$$J = \frac{\pi c^4}{2}. \tag{4.1}$$

The stress distribution is given by

$$\tau = \frac{Tr}{J}. \tag{4.2}$$

4.2 Thin-walled Circular Tube

Consider a tube of outer radius c_o and inner radius c_i. The thickness of the tube t is given by

$$t = c_o - c_i.$$

The mean radius c is defined as

$$c = \frac{c_o + c_i}{2} = c_o - t/2 = c_i + t/2.$$

By *thin-walled* we mean that t/c is negligible compared to unity. The effective value of J for this case is

$$J = \frac{\pi}{2}[c_o^4 - c_i^4] = \frac{\pi}{2}[(c + t/2)^4 - (c - t/2)^4].$$

Using the thin-wall assumption, we can expand the terms using the binomial theorem to get

$$J = \frac{\pi c^4}{2}\left\{\left[1 + 4\left(\frac{t}{2c}\right) + \frac{4 \times 3}{2}\left(\frac{t}{2c}\right)^2 + \cdots\right] - \left[1 - 4\left(\frac{t}{2c}\right) + \frac{4 \times 3}{2}\left(\frac{t}{2c}\right)^2 + \cdots\right]\right\}.$$

This can be simplified by neglecting higher order terms in t/c to get

$$J = 2\pi c^3 t. \tag{4.3}$$

Fig. 4.2 shows the linear stress distribution across the thickness of the tube. When the thickness is small compared to the radius, the variation of the shear stress

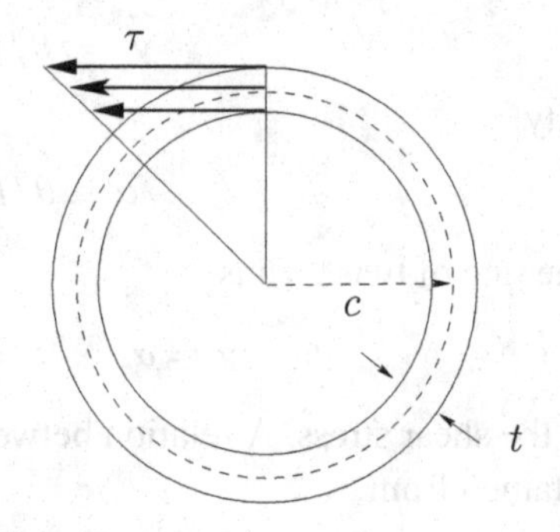

Figure 4.2 Thin-walled circular tube under torsion.

across the thickness is negligible and the mean shear stress corresponds to the mean radius c. The torque created by the mean shear stress τ can be found from

$$T = 2\pi c t \tau c, \quad \tau = \frac{T}{2\pi c^2 t}. \tag{4.4}$$

From this J is found as shown in Eq. (4.3).

4.3 Single-Cell Tube

For a thin-walled closed tube of arbitrary shape the mean surface can be described by two coordinates: z and s. The parameter s is measured from a chosen starting point $s = 0$ and as s is increased it eventually reaches the starting point when $s = s_{max}$. It is the convention that the mean surface is traced in the counterclockwise sense. What follows is known as the Bredt-Batho theory of thin-walled tubes. We need an important assumption at the outset as to the means of maintaining the shape of the cross-section. This is called the Closely Spaced Rigid Diaphragms (CSRD) assumption. We assume that rigid diaphragms (ribs) are inserted inside the tube to maintain its cross-sectional shape. For an airplane wing, the cross-sectional shape represents the selected airfoil and the lift and drag calculations are based on maintaining its shape. Under the CSRD assumption when a torque is applied the entire cross-section would rotate as a rigid body.

We may allow variable thickness as long as the thin-wall assumption is satisfied. With variable thickness, a quantity called shear flow q is very useful. This is defined as

$$q = \tau t. \tag{4.5}$$

Dimensionally q is between a force and a stress, as it has the dimension of force per unit length. As shown in Fig. 4.3, consider the equilibrium of a small element

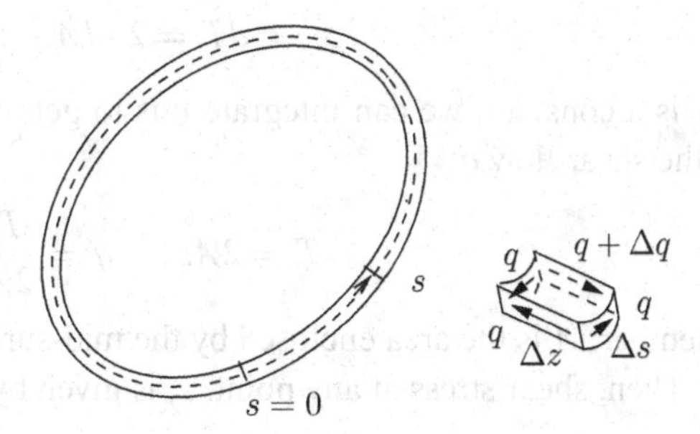

Figure 4.3 Coordinate s and the force balance of an element of a thin-walled arbitrary cross-section tube.

of length Δz and width Δs. If q varies with s, one side has a force of $q\Delta z$ and the other side $(q + \Delta q)\Delta z$, where

$$\Delta q = \frac{\partial q}{\partial s}\Delta s.$$

As there is no net force in the z-direction, $\partial q/\partial s = 0$. Thus, q is a constant along s, even when the thickness changes. Where the thickness is small there is higher shear stress and vice versa. This is analogous to a steady state, incompressible fluid flow in a pipe, where the flow rate is constant. Fluid speed increases when pipe cross-sectional area decreases. This flow analogy suggests the term "shear flow."

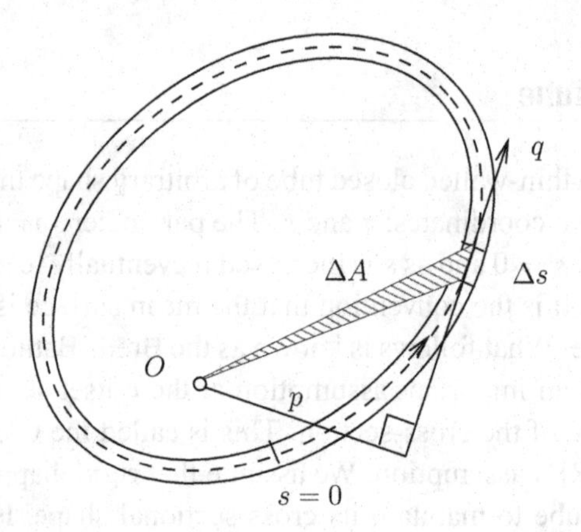

Figure 4.4 Evaluation of torque due to the shear flow.

Fig. 4.4 shows a small element Δs with a shear flow q. The torque produced by the force $q\,\Delta s$ about an arbitrary point O is given by

$$\Delta T = pq\Delta s, \tag{4.6}$$

where p is the perpendicular distance from O to the line of action of the force. From the hatched wedge, it can be seen that the area of the wedge is

$$\Delta A = p\Delta s/2. \tag{4.7}$$

Using this,

$$dT = 2q\,dA. \tag{4.8}$$

As q is a constant, we can integrate this to get a relation between the torque T and the shear flow q,

$$T = 2Aq, \quad q = \frac{T}{2A}. \tag{4.9}$$

Remember: A is the area enclosed by the mid-surface curve and not the material area. Then, shear stress at any point, s, is given by

$$\tau = \frac{T}{2At}. \tag{4.10}$$

To complete this theory, we would like to find the rate of twist $\alpha = \theta/L$ in terms of T. We use the complementary energy to do this. The energy stored in an element ds along the length L of the tube is

$$dU^* = \frac{\tau^2 tLds}{2G} = \frac{q^2 Lds}{2tG} = \frac{T^2 ds}{8A^2 tG}.$$

Integrating this, we find

$$U^* = \frac{T^2 L}{8A^2} \oint \frac{ds}{tG},$$

where the integral has to be taken around the tube. As this formula stands, variable t and variable G can be allowed. Although it is common to find variable t, variable G is rare. Taking G outside of the integral and differentiating with T, we have

$$\theta = \frac{TL\delta}{4A^2G}, \quad \alpha = \frac{T\delta}{4A^2G}, \tag{4.11}$$

where

$$\delta = \oint \frac{ds}{t}. \tag{4.12}$$

4.3.1 Example: A Rectangular Tube

Consider a tube of rectangular cross-section with height $2a$, width $4a$, and wall thickness t. For this $A = 8a^2$. Then the shear flow is given by

$$q = \frac{T}{2A} = \frac{T}{16a^2} = 0.0625\frac{T}{a^2}.$$

The integral

$$\delta = \oint \frac{ds}{t} = \frac{12a}{t}.$$

The rate of twist is obtained from

$$\alpha = \frac{T\delta}{4A^2G} = \frac{3T}{64Ga^3t}.$$

4.4 Multi-cell Tubes

Fig. 4.5 shows a sketch of a three-cell tube. The cells are numbered as 1, 2, 3, ... from left to right. We use the analogy of denoting electrical currents in a circuit to

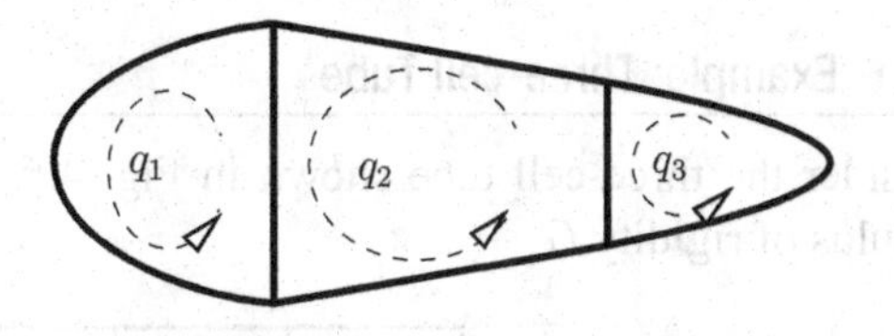

Figure 4.5 Shear flows in a three-cell tube.

indicate the shear flows $q_1, q_2, q_3, \ldots$. For example, in cell 2 the bottom and top walls have shear flows q_2, the left wall has $(q_2 - q_1)$, and the right wall $(q_2 - q_3)$. The net torque produced by all three shear flows is given by

$$T = 2A_1q_1 + 2A_2q_2 + 2A_3q_3, \tag{4.13}$$

where A_i is the enclosed area of the ith cell.

This gives one equation involving three unknowns: $q_i, i = 1, 2, 3$. Compatibility requires that all three cells rotate equally. We may find the angle of rotation of a particular cell, say the ith, using the complementary virtual work. Let us apply a virtual torque δT on this cell, which produces a virtual shear flow δq around the ith cell, with

$$\delta q = \frac{\delta T}{2A_i}. \tag{4.14}$$

Equating the complementary virtual work with the change in stored energy we get

$$\delta T\theta = L\oint_i \frac{q\delta q}{Gt^2} t ds = L\frac{\delta T}{2A_i}\oint_i \frac{q}{Gt} ds. \tag{4.15}$$

This can be written as

$$\alpha = \frac{1}{2GA_i}\oint_i \frac{q ds}{t}. \tag{4.16}$$

From this, noting that q_i goes around the cell and q_{i-1} opposes it on the left wall and q_{i+1} on the right wall, we may expand Eq. (4.16) as

$$\alpha = \frac{1}{2GA_i}\left[q_i\oint_i \frac{ds}{t} - q_{i-1}\int_{Li}\frac{ds}{t} - q_{i+1}\int_{Ri}\frac{ds}{t}\right], \tag{4.17}$$

where $\oint_i$ has to be evaluated around the ith cell, $\int_{Li}$ and $\int_{Ri}$ are evaluated over the left and right walls of the cell. We may use the notations

$$\delta_i = \oint_i \frac{ds}{t}, \quad \delta_{Li} = \int_{Li}\frac{ds}{t}, \quad \delta_{Ri} = \int_{Ri}\frac{ds}{t}, \tag{4.18}$$

to write the equations for the rate of twist for the three cells,

$$\frac{1}{2GA_i}[\delta_i q_i - \delta_{Li}q_{i-1} - \delta_{Ri}q_{i+1}] = \alpha, \quad i = 1, 2, 3. \tag{4.19}$$

In the case of variable G, it has to be kept inside the integrals. Eqs. (4.13) and (4.19) form four equations for fours unknowns q_i and α. Also, note that the leftmost cell has no left wall and the rightmost cell has no right wall.

4.4.1 Example: Three-cell Tube

Consider the three-cell tube shown in Fig. 4.6. All walls have thickness t and modulus of rigidity G.

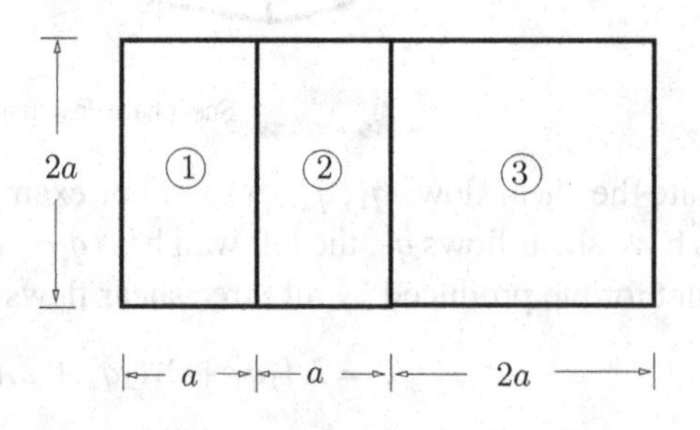

Figure 4.6 A three-cell tube with uniform thickness t.

We assume a torque T acts on it. Find shear flows in the walls and the rate of twist.

The cell areas are

$$A_1 = 2a^2, \quad A_2 = 2a^2, \quad A_3 = 4a^2.$$

With shear flows q_1, q_2, and q_3, the torque–shear flow relation is

$$2A_1q_1 + 2A_2q_2 + 2A_3q_3 = T,$$

$$\text{i.e.} \quad q_1 + q_2 + 2q_3 = \frac{T}{4a^2}. \tag{4.20}$$

The rate of twist is given by

$$\delta_i q_i - \delta_{Li} q_{i-1} - \delta_{Ri} q_{i+1} = 2GA_i\alpha, \quad i = 1, 2, 3.$$

For this case

$$\delta_1 = 6a/t, \quad \delta_2 = 6a/t, \quad \delta_3 = 8a/t,$$

$$\delta_{L1} = 0, \quad \delta_{R1} = \delta_{L2} = 2a/t, \quad \delta_{R2} = \delta_{L3} = 2a/t, \quad \delta_{R3} = 0.$$

Thus

$$3q_1 - q_2 = 2Gat\alpha,$$

$$-q_1 + 3q_2 - q_3 = 2Gat\alpha,$$

$$-q_2 + 4q_3 = 4Gat\alpha.$$

Solving the three equations we get

$$q_1 = 1.1724Gat\alpha, \quad q_2 = 1.5117Gat\alpha, \quad q_3 = 1.3793Gat\alpha.$$

The rate twist α is found from the torque relation, Eq. (4.20), as

$$3.3427Gat\alpha = \frac{T}{4a^2}, \quad \text{or} \quad \alpha = 0.0459\frac{T}{Ga^3t}.$$

Then

$$(q_1, q_2, q_3) = (0.0538, 0.0696, 0.0630)\frac{T}{a^2}.$$

We note that the walls separating the cells have a much smaller amount of shear flows compared to the top and bottom walls. The shear flows in the top and bottom walls are comparable to the value we obtained for a single cell tube of the same outer dimensions in the previous example.

4.5 Open Thin-walled Tubes

Open-section tubes are really not tubes, but thin sheets, as shown in Fig. 4.7. When a torque T is applied on this cross-section, shear stresses, τ, develop, creating an equivalent torque. The material in the positive η-direction has shear stresses pointing in the negative s-direction and vice versa. The simplest assumption we can make is that τ is linearly distributed as shown in 4.7(a), varying from 0 to a

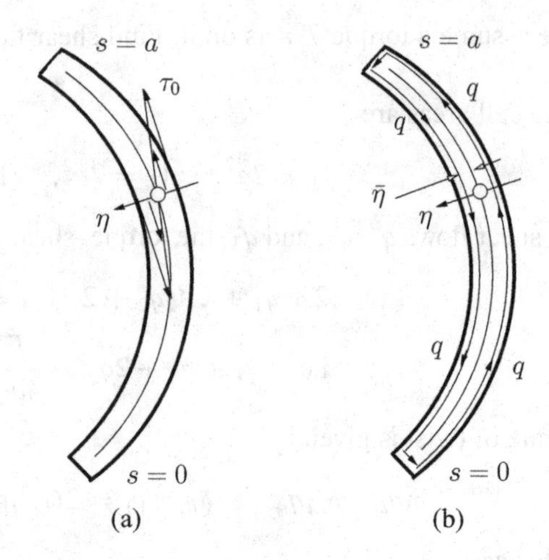

Figure 4.7 Open cross-section with assumed shear stress distribution.

maximum of τ_0. There is no shear stress on the mid-surface, $\eta = 0$. Effectively, we may think of this sheet as a single-cell tube with a hole of zero area at the mid-surface. Fig. 4.7(b) shows a shear flow q at a distance $\bar{\eta}$ going around this tube. Equating the force per unit length due to the shear stress to q, we get

$$q = \frac{1}{2}\tau_0\frac{t}{2}, \quad \text{or} \quad \tau_0 = 4q/t. \tag{4.21}$$

Now, we have to decide the location, $\bar{\eta}$, of the line of action of q. For this we equate the moment produced by the linearly distributed shear stress to that due to q. That is

$$\bar{\eta}q = \frac{2}{3}\frac{t}{2}\frac{\tau_0 t}{4} = \frac{1}{3}tq, \quad \text{or} \quad \bar{\eta} = \frac{t}{3}. \tag{4.22}$$

The area enclosed by our fictitious tube is

$$A = 2at/3. \tag{4.23}$$

This gives

$$q = \frac{T}{2A} = \frac{3T}{4at}, \quad \tau_0 = \frac{3T}{at^2}. \tag{4.24}$$

To find the rate of twist we use the energy method.

$$U^* = \frac{L}{2G}\int_0^a \int_{-t/2}^{t/2} \tau^2 d\eta ds. \tag{4.25}$$

Here, from the linear stress distribution,

$$\tau = 2\tau_0\eta/t. \tag{4.26}$$

Then

$$U^* = \frac{4La}{G}\frac{\tau_0^2}{t^2}\int_0^{t/2}\eta^2 d\eta,$$
$$= \frac{4La}{G}\frac{9T^2}{a^2t^4}\frac{t^3}{24},$$
$$= \frac{3LT^2}{2Gat^3}.$$

Using $dU^*/dT = \theta$ and $\alpha = \theta/L$, we get

$$\alpha = \frac{T}{GJ}, \tag{4.27}$$

where

$$J = \frac{at^3}{3}. \tag{4.28}$$

As we may expect, the stiffness of the open tube is many orders of magnitude smaller than that of a closed tube.

4.5.1 Example: Circular Tube

Consider a circular single-cell tube of radius a and thickness t. For an applied torque T, the shear flow and shear stress are given by

$$q = \frac{T}{2\pi a^2}, \quad \tau = \frac{T}{2\pi a^2 t}.$$

In the case where this tube develops a crack along its length, we have an open tube. The maximum shear stress under the same applied torque is

$$\tau_{open} = \frac{3T}{2\pi at^2},$$

which is (a/t) times larger than the value for the closed tube. The torsional stiffness for the uncracked tube is

$$GJ = 4GA^2/\delta = 4G\pi^2a^4/(2\pi a/t) = 2\pi Ga^3t.$$

For the open tube

$$GJ_{open} = \frac{G}{3}2\pi at^3 = 2\pi Gat^3/3,$$

which is of the order of $(t/a)^2$ smaller.

4.6 Warping of the Cross Section

In the case of noncircular tubes, plane cross-sections before deformation do not remain plane after. This phenomenon is called warping. We assume we are dealing with a particular cell (say, i^{th}) of a multicell tube. In order to describe warping, we introduce displacement components w along z and u tangential to the mid-surface at s. We complete this set by adding a coordinate η and a displacement v

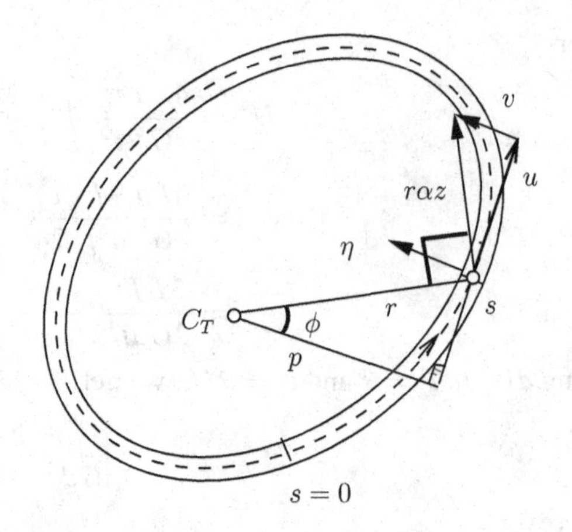

Figure 4.8 Tangential and normal displacements around the tube.

perpendicular to the tangent as shown in Fig. 4.8. The displacements are exaggerated in this figure for clarity. The shear stress τ, going around the cell, can be expressed as

$$\tau = \sigma_{zs}.$$

The corresponding shear strain has the form

$$\gamma_{zs} = \frac{\partial w}{\partial s} + \frac{\partial u}{\partial z} = \tau/G. \tag{4.29}$$

The relative rigid body rotations of the tube take place around a fixed point called the *center of twist*, C_T. At this stage of the development of our theory we don't know how to find this point. Consider a cross-section at z that has undergone a rigid body rotation of αz relative to the section at $z = 0$. Let a point at s be at a distance r from the center of twist C_T. This point will move perpendicular to the radius r by αzr. This displacement can be resolved to have the tangential component

$$u = \alpha zr \cos\phi = \alpha zp, \tag{4.30}$$

where p is the perpendicular distance from the center of twist to the tangent at s as shown in Fig. 4.8. This gives

$$\frac{\partial u}{\partial z} = \alpha p. \tag{4.31}$$

From the equation for the shear strain, this gives

$$\frac{\partial w}{\partial s} = \frac{q}{tG} - p\alpha. \tag{4.32}$$

Assuming G is constant, we integrate this expression form $s = 0$ to s, keeping in mind, in the case of a multicell tube, q is a variable (piecewise constant) depending on s, to get

$$w(s) = w_0 + \frac{1}{G}\oint_i \frac{q\,ds}{t} - \alpha\int_0^s p\,ds, \tag{4.33}$$

where w_0 is the axial displacement at our origin $s = 0$.

If we evaluate $w(s)$ at the end point $s = s_{max}$, we should get w_0. This leads to

$$\alpha = \frac{1}{2GA_i} \oint_i \frac{q ds}{t}, \tag{4.34}$$

which is in agreement with the formula for the rate of twist, α, derived earlier in Eq. (4.16) using virtual work.

We may specialize the formula for the warp distribution in Eq. (4.33) for a single-cell tube by noting q is a constant given by

$$q = \frac{T}{2A}. \tag{4.35}$$

Introducing the notations

$$\delta_s = \int_0^s \frac{ds}{t}, \quad A_s = \int_0^s \frac{p ds}{2}, \tag{4.36}$$

we have the expression for warping displacement,

$$w(s) - w_0 = \frac{T}{2A^2 G} \left[A\delta_s - A_s \delta \right]. \tag{4.37}$$

Clearly, the right-hand side of this equation vanishes when s reaches s_{max}. The warping of the section represented by $w(s)$ is called the *primary warping* as in contrast to the warping perpendicular to the mid-surface, in the η-direction. With our assumption of thin-walled tubes the *secondary warping* in the η-direction is negligible.

Once again, it has to be kept in mind that the area, $A_s = \int_0^s pds/2$, uses the perpendicular distance p from a special point, namely, the center of twist. When a tube cross-section has two perpendicular axes of symmetry (bilateral symmetry), we can argue that the center of twist lies at the intersection of these axes. Further, the warping displacement has to be zero at points of symmetry on the walls.

4.6.1 Example: Warping in a Single-cell Tube

Consider a single-cell symmetric tube of height $2a$ and width $2b$ with thicknesses t_a for the vertical walls and t_b for the horizontal walls, shown in Fig. 4.9.

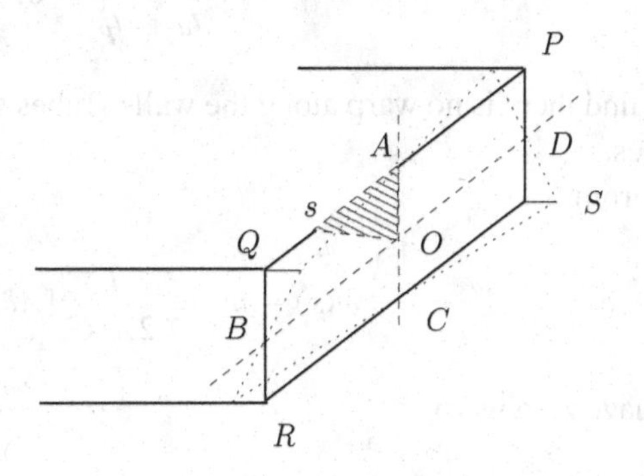

Figure 4.9 Warping of the walls in a rectangular tube.

The planes AC and BD form symmetry planes and their intersection O is the center of twist. The points A, B, C, and D are symmetry points where the warp is zero. The area of the cell is $4ab$. If we measure s from the symmetry point A, then, with $w_A = 0$,

$$\begin{aligned} w(s) &= \frac{T}{2A^2G}\left[A\delta_s - A_s\delta\right], \\ &= \frac{T}{32a^2b^2G}\left[4ab\frac{s}{t_b} - \frac{sa}{2}\left(\frac{4a}{t_a} + \frac{4b}{t_b}\right)\right], \\ &= \frac{T}{16a^2b^2G}\left[\frac{b}{t_b} - \frac{a}{t_a}\right] as. \end{aligned}$$

Thus, the magnitude of the warp increases linearly from A to Q. At Q, we have

$$w_Q = \frac{T}{16abG}\left[\frac{b}{t_b} - \frac{a}{t_a}\right].$$

To compute the warp distribution from Q to B, we may use a new coordinate s starting at Q. Then

$$\begin{aligned} w(s) = w_Q + &\frac{T}{32a^2b^2G}\left[4ab\frac{s}{t_a} - \frac{sb}{2}\left(\frac{4a}{t_a} + \frac{4b}{t_b}\right)\right], \\ &= \frac{T}{16a^2b^2G}\left(\frac{b}{t_b} - \frac{a}{t_a}\right) b(a - s). \end{aligned}$$

At the point B, we get $w_B = 0$. If we extend s to R, we find

$$w_R = -w_Q.$$

Thus, the warp distribution is antisymmetric about the symmetry points and the over all warp distribution is as shown by the dotted line in Fig. 4.9. If

$$\frac{a}{t_a} - \frac{b}{t_b} = 0,$$

we find there is no warp along the walls. Tubes with zero warp are called Neuber tubes.

From

$$w(s) - w_0 = \frac{T}{2A^2G}\left[A\delta_s - A_s\delta\right],$$

to have zero warp

$$\delta_s = \frac{\delta}{A}A_s.$$

This can be expressed as

$$\int_0^s \frac{ds}{t} = \frac{\delta}{2A}\int_0^s pds.$$

Differentiating, we find

$$\frac{1}{t} = \frac{\delta}{2A}p, \quad \text{or} \quad pt = 2A/\delta,$$

as a general condition for zero warp. Of course, in the case of variable G, it has to be kept inside the integrals.

4.6.2 Warping in Open Tubes

The major difference between open and closed tubes is that the shear stress along the mid-surface is zero in the former and non-zero in the latter. In our formula for warping displacement, Eq. (4.33), the first term is due to the shear stress. In the open sections we drop this term and what remains is

$$w(s) = w(0) - \alpha \int_0^s pds = w(0) - 2\alpha A_s. \tag{4.38}$$

4.6.3 Example: Split Circular Tube

Consider an open tube of radius a obtained by introducing a small cut between A and B, as shown in Fig. 4.10. This section has one plane of symmetry and the center of twist lies on this. In fact, it is outside the tube at point O, as we will see

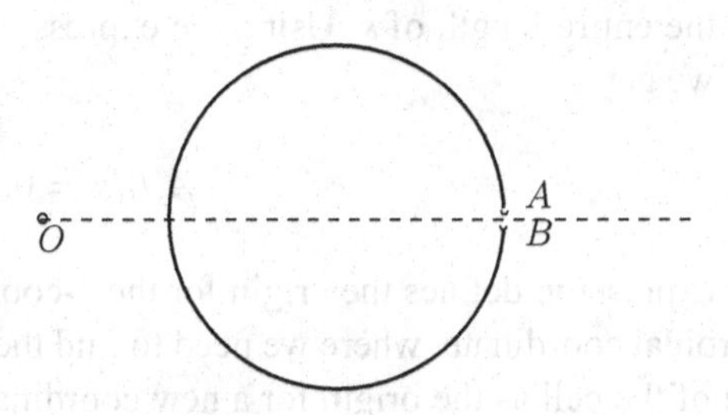

Figure 4.10 Warping in a split circular tube.

in the next chapter. If we are only interested in the relative warp between A and B, we use the relation

$$w_B - w_A = -\alpha \int_0^{2\pi a} pds = -\alpha 2A = -2\pi\alpha a^2.$$

Our treatment of the tube is based on freely allowing warping of the walls. If warping is prevented (say, by fixing one end of the tube to a wall), normal stress σ_{zz} would develop from this strain. We will study this in detail in the next section.

4.7 Induced Torque due to Varying Warp

As we have done before, we use the z- coordinate along the length of the tube and an s-coordinate around the tube. The origin of the s-coordinate is not arbitrary; it has to start from a point where warp is zero. When the end $z = 0$ is built into a wall, warping of that end is prevented. Along any line $s =$ constant, the warp will vary along z. We rewrite Eq. (4.38) as

$$w(s, z) = w(0, z) - 2\frac{d\theta}{dz}A_s, \tag{4.39}$$

to reflect the axial variation of the warp. If we choose as the origin for the s-coordinate the point where $w = 0$,

$$w(s, z) = -2\frac{d\theta}{dz}A_s. \tag{4.40}$$

The axial strain ϵ_{zz} is obtained as

$$\epsilon_{zz} = -2\frac{d^2\theta}{dz^2}A_s, \tag{4.41}$$

and the corresponding axial stress is

$$\sigma_{zz} = -2E\frac{d^2\theta}{dz^2}A_s. \tag{4.42}$$

As the net axial force applied to the tube is zero,

$$\int_e \sigma_{zz} t ds = 0, \tag{4.43}$$

where the subscript e indicates that the integral has to be taken from edge to edge, over the entire length of s. Using the expression (4.42), where θ is independent of s, we get

$$\int_e A_s t ds = 0. \tag{4.44}$$

This expression defines the origin for the s-coordinate. This is similar to using a centroidal coordinate, where we need to find the centroid first. If we start with an edge of the cell as the origin for a new coordinate s', we can write

$$A_s = A_0 + \int_0^s p' ds'/2 = A_0 + A'_s. \tag{4.45}$$

Then, the zero-force condition gives

$$A_0 \int_e t ds' + \int_e A'_s t ds' = 0, \quad A_0 = -\frac{\int_e A'_s t ds'}{\int_e t ds'}. \tag{4.46}$$

There is a point s^* from the edge which, when used in Eq. (4.45), makes $A_s = 0$. That is

$$\int_0^{s*} p' ds'/2 = -A_0. \tag{4.47}$$

This establishes $s = s' - s^*$. Using the stress in Eq. (4.42) in the equilibrium equation expressed in the s-coordinate,

$$t\frac{\partial \sigma_{zz}}{\partial z} + \frac{\partial q}{\partial s} = 0, \tag{4.48}$$

we can integrate from an edge of the cell where $q = 0$, to have

$$q = -\int_0^s t\frac{\partial \sigma_{zz}}{\partial z} ds. \tag{4.49}$$

Substituting for σ_{zz},

$$q = 2E\frac{d^3\theta}{dz^3}\int_0^s A_s t ds. \tag{4.50}$$

This creates an induced torque due to the constrained warp,

$$T_w = \int_e pq ds = 2E\frac{d^3\theta}{dz^3}\int_e p\int_0^s A_s t ds\, ds. \tag{4.51}$$

The double integral can be simplified using integration by parts.

$$\int_e p\int_0^S A_s t ds\, ds = 2\int_e \frac{dA_s}{ds}\int_0^S A_s t ds\, ds = 2\left[A_s\int_0^S 2A_s t ds\right]_e - \int_e 2A_s^2 t ds. \tag{4.52}$$

The first term in this expression is zero because q at the edge given by the integral inside is zero. Then

$$T_w = -E\Gamma\frac{d^3\theta}{dz^3}, \tag{4.53}$$

where Γ is the torsional-bending constant given by

$$\Gamma = 4\int_e A_s^2 t ds. \tag{4.54}$$

Part of the applied external torque T produces this induced torque T_w and the remaining part creates the rate of twist $d\theta/dz$. That is

$$T = GJ\frac{d\theta}{dz} - E\Gamma\frac{d^3\theta}{dz^3}. \tag{4.55}$$

This third-order differential equation needs three boundary conditions. The fixed end where warp is prevented will have

$$\theta = 0, \quad \frac{d\theta}{dz} = 0. \tag{4.56}$$

The other end where warp is allowed will have $\sigma_{zz} = 0$, or

$$\frac{d^2\theta}{dz^2} = 0. \tag{4.57}$$

This theory is known as the Wagner theory of torsion.

4.7.1 Example: Channel Section

Consider a channel section with flange width a, web depth $2a$, and length L, shown in Fig. 4.11. It is known that the center of twist for this cross-section is

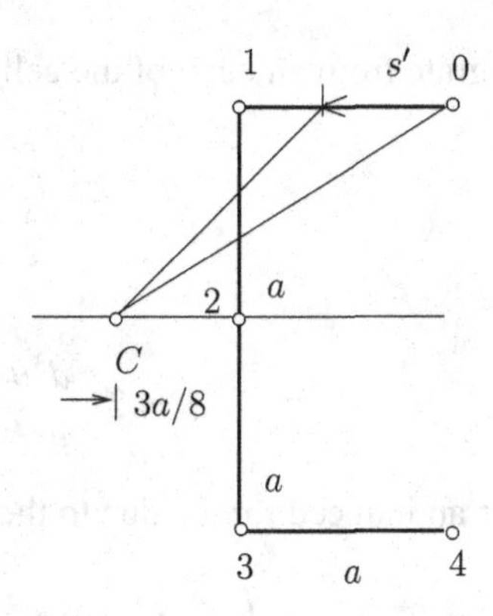

Figure 4.11 A channel section with warp constraint.

located at a distance $3a/8$ to the left of the web. Although we know from symmetry that the point of zero warp is located at the point 2, we will proceed as if we don't know this. We start with s' measured from point 0. The area calculations involve areas of triangles with known base and altitude. From $0 \to 1$,

$$A'_s = \frac{as'}{2}, \quad A'_1 = \frac{a^2}{2}, \tag{4.58}$$

where the subscript 1 indicates the location. As we move along from 0 to 1, the angle of the line from C to s' increases. After 1, the angle decreases, contributing negative area. Using a new s' starting from 1, in the section $1 \to 3$,

$$A'_s = \frac{a^2}{2} - \frac{3as'}{16}, \quad A'_2 = \frac{5a^2}{16}, \quad A'_3 = \frac{a^2}{8}. \tag{4.59}$$

To go from 3 to 4, we use another s' starting from 3. From $3 \to 4$,

$$A'_s = \frac{a^2}{8} + \frac{as'}{2}, \quad A'_4 = \frac{5a^2}{8}. \tag{4.60}$$

Using the formula, (4.46),

$$A_0 = -\frac{\int_e A'_s t ds'}{\int_e t ds'}, \tag{4.61}$$

we get

$$\begin{aligned} A_0 &= -\frac{1}{4a}\left[\int_0^a \frac{as'}{2} ds' + \int_0^{2a} \frac{(8a^2 - 3as')}{16} ds' + \int_0^a \frac{(a^2 + 4as')}{8} ds'\right], \\ &= -\frac{5a^2}{16}. \end{aligned} \tag{4.62}$$

Then $A_s = A_0 + A'_s$ becomes zero at 2, the point of symmetry. Now, including A_0,

$$0 \to 1, \quad A_s = \frac{as'}{2} - \frac{5a^2}{16},$$

$$1 \to 3, \quad A_s = \frac{3a^2 - 3as'}{16},$$

$$3 \to 4, \quad A_s = \frac{as'}{2} - \frac{3a^2}{16}.$$

Using Eq. (4.54),

$$\begin{aligned}\Gamma &= 4\int_e A_s^2 t ds,\\ &= \frac{4t}{256}\left[\int_0^a (8as' - 5a^2)^2 ds' + \int_0^{2a} 9(a^2 - as')^2 ds' + \int_0^a (8as' - 3a^2)^2 ds\right],\\ &= \frac{7a^5 t}{24}.\end{aligned} \tag{4.63}$$

The applied torque is related to the rate of twist in the form

$$GJ\frac{d\theta}{dz} - E\Gamma\frac{d^3\theta}{dz^3} = T. \tag{4.64}$$

This can be written as

$$\frac{d^3\theta}{dz^3} - \mu^2\frac{d\theta}{dz} = -\frac{T}{E\Gamma}, \tag{4.65}$$

where

$$\mu^2 = \frac{GJ}{E\Gamma}. \tag{4.66}$$

For this open section

$$J = 4at^3/3, \quad \mu = \frac{8}{\sqrt{5}}\frac{t}{a^2}. \tag{4.67}$$

Then,

$$\mu = \sqrt{\frac{32Gt^2}{7Ea^4}}. \tag{4.68}$$

The solution of the differential equation can be written as

$$\frac{d\theta}{dz} = \frac{T}{GJ}[1 + A\sinh\mu z + B\cosh\mu z], \tag{4.69}$$

where A and B are found using

$$\frac{d\theta}{dz} = 0, \quad \text{at} \quad z = 0, \tag{4.70}$$

and

$$\frac{d^2\theta}{dz^2} = 0, \quad \text{at} \quad z = L, \tag{4.71}$$

as

$$B = -1, \quad A = -\tanh\mu L. \tag{4.72}$$

The final form can be simplified to read

$$\frac{d\theta}{dz} = \frac{T}{GJ}\left[1 - \frac{\cosh\mu(L-z)}{\cosh\mu L}\right]. \tag{4.73}$$

The quantity μL is of the order of tL/a^2, which is neither very large nor very small. This indicates that the effect of warping constraint and the accompanying axial stress can extend to a substantial part of the tube.

4.7.2 Example: Split Circular Tube

For a split circular tube with radius R, thickness t, and length L (shown in Fig. 4.12), it will be shown that the center of twist C is located at a distance $2R$

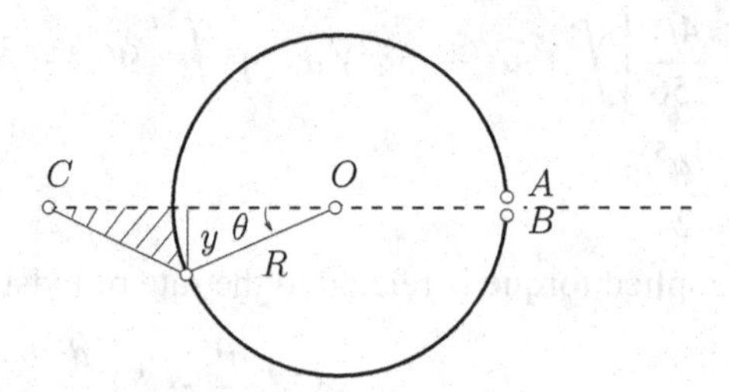

Figure 4.12 A split circular tube subjected to a torque.

from the center. To compute torsional-bending constant Γ, we vary θ from $-\pi$ to π, with $\theta = 0$ being the point of symmetry where the warp is zero. Then

$$A_s = R^2 \sin\theta - \frac{1}{2}R^2\theta. \tag{4.74}$$

The torsional-bending constant Γ can be found from

$$\begin{aligned}
\Gamma &= 4t\int_{-\pi}^{\pi} A_s^2 R d\theta = 8tR^5 \int_0^{\pi}\left[\sin\theta - \frac{\theta}{2}\right]^2 d\theta,\\
&= 8tR^5 \int_0^{\pi}\left[\sin^2\theta - \theta\sin\theta + \frac{\theta^2}{4}\right] d\theta,\\
&= 8tR^5\left[\frac{\pi}{2} + (\theta\cos\theta - \sin\theta)\Big|_0^{\pi} + \frac{\pi^3}{12}\right],\\
&= \frac{2tR^5}{3}[\pi^3 - 6\pi].
\end{aligned} \tag{4.75}$$

Note that the dimension of Γ is length raised to the sixth power.

FURTHER READING

Curtis, H. D., *Fundamentals of Aircraft Structural Analysis*, Irwin (1997).
Megson, T. H. G., *Aircraft Structures for Engineering Students*, Butterworth-Heinemann (2007).
Rivello, R. M., *Theory and Analysis of Flight Structures*, McGraw-Hill (1969).
Sun, C. T., *Mechanics of Aircraft Structures*, John Wiley (2006).

EXERCISES

4.1 Consider a 10 cm mean radius hollow tube with thickness 2 mm. Using the exact value of J and the thin-wall approximation, $J = 2\pi r^3 t$ and $\tau = T/(2At)$, obtain the percentage error in the maximum shear stress and in the rate of twist.

4.2 Three tubes are fabricated with thickness t and mean surface length a around the tubes, in the form of a circle, an equilateral triangle, and a square. Using the thin wall approximation, compare the shear stresses and the torsional stiffnesses of these three shapes with the circle as the reference.

4.3 A two-cell tube has outer walls with thickness $2t$ and an inner wall with thickness t (see Fig. 4.13). Calculate the torsional stiffness of this tube if the modulus of rigidity is G. What are the shear flows in the different walls? Number the cells from left to right.

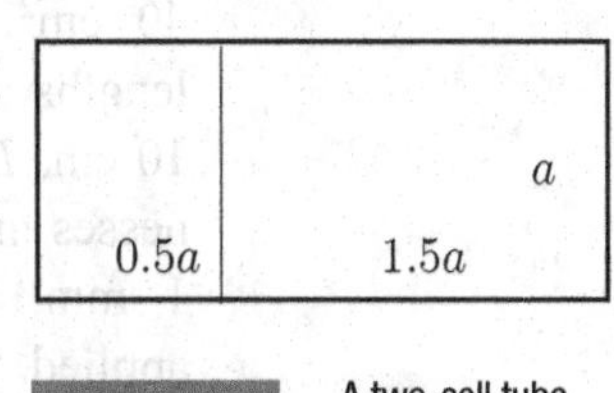

Figure 4.13 A two-cell tube.

4.4 For the two-cell tube shown in Fig. 4.13, the rate of twist is given as 5° per m and $G = 20$ GPa. If $a = 10$ cm and $t = 4$ mm, compute the maximum shear stress in the walls.

4.5 Two shafts of equal length are joined at the middle. The shaft on the left has a solid circular cross-section of radius 5 cm and the one on the right has a hollow section with inner radius 4.8 cm and outer radius 5 cm. What are the maximum shear stresses and rates of twist in the two sections if the applied torque is 100 Nm and $G = 20$ GPa? The torque is applied at the point where the two shafts meet.

4.6 A three-cell tube is shown in Fig. 4.14. The outer walls have a thickness of $2t$ and the inner walls t. Compute the shear flows in the different walls and find the torsional stiffness assuming homogeneous material properties.

4.7 A torque of 100 Nm acts on tube shown in Fig. 4.15. The dimensions of the tube are $a = 10$ cm and the thickness is uniform with a value of 5 mm. The modulus of rigidity of the material is 20 GPa. Compute the shear flows in all the walls and the rate of twist.

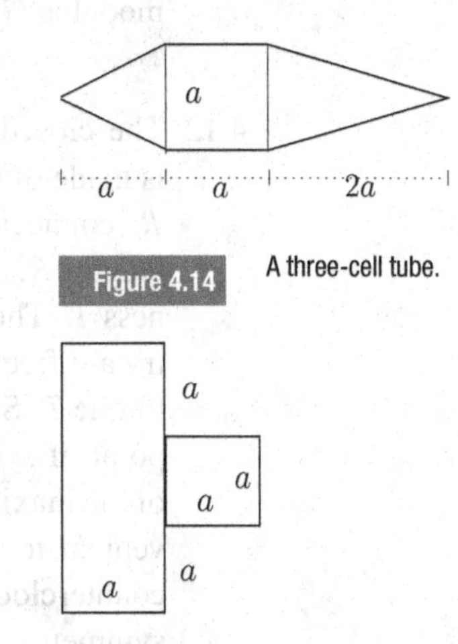

Figure 4.14 A three-cell tube.

Figure 4.15 A two-cell tube.

4.8 For the slit circular cell shown in Fig. 4.16, calculate the warp distribution $w(\phi)$ in terms of the applied torque T, the thickness t, and the radius R. The center of twist is known to be at a distance $2R$ to the left of the center of the circle.

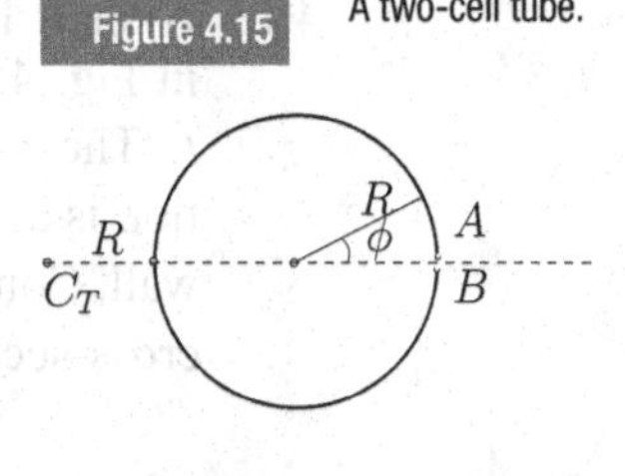

Figure 4.16 A circular cell with a slit.

4.9 The center of twist of the channel section given in Fig. 4.17 is at $e = 16a^4t/I$ to the left of the vertical web on the line of symmetry. Compute the warps at the two edges of the flanges in terms of the applied torque T, the shear modulus G, and the thickness t.

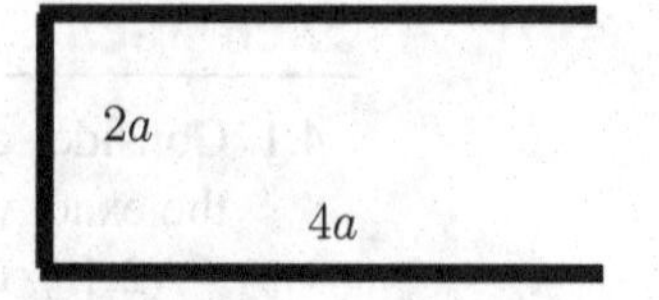

Figure 4.17 A channel section tube.

4.10 For the two-cell tube shown in Fig. 4.18, the cell areas are: $A_1 = 40\ \text{cm}^2$, $A_2 = 60\ \text{cm}^2$. The wall lengths are: $L_{12} = 13$ cm, $L_{13} = 10$ cm, $L_{34} = 8$ cm. The wall thicknesses are: $t_{12} = t_{23} = 2$ mm, $t_{13} = 1$ mm, $t_{34} = t_{41} = 3$ mm. If the applied torque is 200 Nm and the modulus of rigidity is 20 GPa, compute the shear stresses in the walls and the rate of twist.

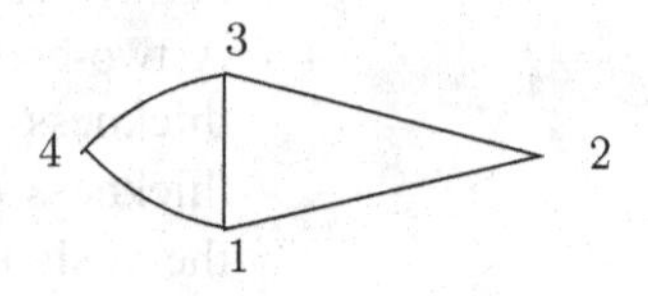

Figure 4.18 A two-cell tube.

4.11 For the channel section shown in Fig. 4.19, the center of twist is located at $e = 3a/8$ left of the vertical wall. Compute the relative warp $w_A - w_B$ in terms of the applied torque T, the length a, the thickness t, and the shear modulus G.

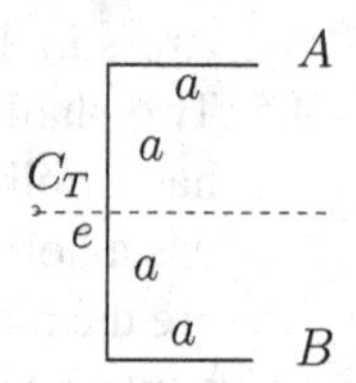

Figure 4.19 A channel section.

4.12 The closed tube shown in Fig. 4.20 is made of two semi-circles of radius R connected by straight lines of length R. It has a uniform thickness t. The four points of symmetry are free of warp under an applied torque T. Starting from the rightmost point of symmetry, find the location of the maximum warp. It will be convenient to use an angle θ measured counterclockwise from the point of symmetry.

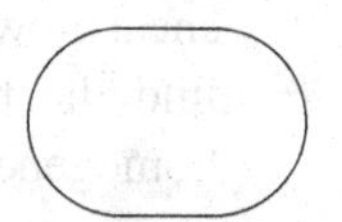

Figure 4.20 A closed section tube.

4.13 A shaft has the cross-section shown in Fig. 4.21, with uniform thickness t. The center of twist for this section is at the midpoint of the vertical wall. Compute I_{xx}, I_{yy}, and Γ for this cross-section.

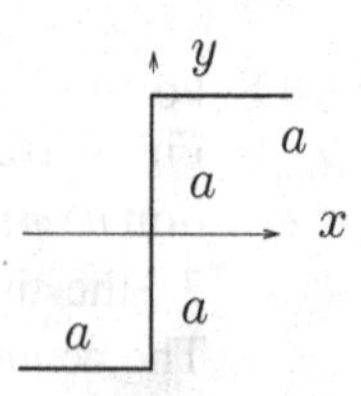

Figure 4.21 A Z-section shaft.

4.14 A shaft of length 2 m has the cross-section shown in Fig. 4.21, with $a = 4$ cm and $t = 4$ mm. Under an applied torque of 200 Nm, calculate the relative angle of rotation between the two ends of the shaft, if warping is prevented at both ends. Assume $E = 200$ GPa and $G = 80$ GPa.

4.15 A shaft of length 2 m has the cross-section shown in Fig. 4.19, with $a = 4$ cm and $t = 4$ mm. Calculate Γ for the shaft. Under an applied torque of 200 Nm, calculate the relative angle of rotation between the two ends of the shaft, if warping is only allowed at one end. Assume the material is isotropic with $E = 200$ GPa and $\nu = 0.3$.

4.16 A shaft has the cross-section shown in Fig. 4.22, with uniform thickness t. The center of twist for this section is at the midpoint of the vertical wall. Compute I_{xx}, I_{yy}, and Γ for this cross-section.

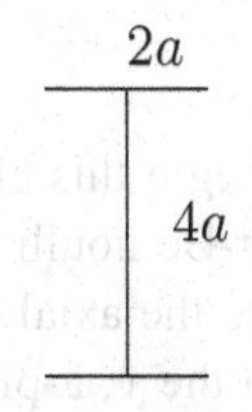

Figure 4.22 An I-section shaft.

4.17 A shaft of length 2 m has the cross-section shown in Fig. 4.22, with $a = 4$ cm and $t = 4$ mm. If warping is prevented at the left end of the shaft and it is allowed on the right end, obtain the rate of twist, α, as function of the axial coordinate z. Use $E/G = 2.6$, $G = 20$ GPa, and the applied torque $T = 200$ Nm.

5 Bending

We begin this chapter by reviewing the elementary beam theory known as the Euler-Bernoulli theory. The beam axis, x, will be placed along the neutral axis where the axial strain due to bending is zero. A cross-section of the beam will be in the y, z-plane. The deflections of the beam in the y and z directions will be denoted by v and w, respectively. A sketch of the undeformed and deformed beam with its projections on the xz-plane and xy-plane is shown in Fig. 5.1. The term "flexure" is also used to describe bending.

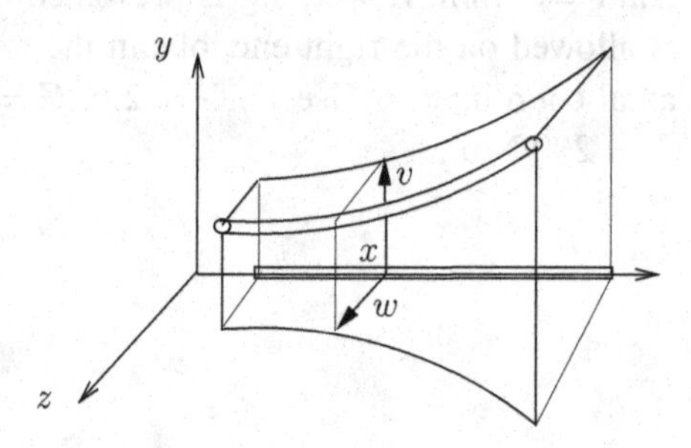

Figure 5.1 Beam deflections in two directions.

The basic assumptions of the elementary beam theory are: (a) plane sections normal to the neutral axis remain plane and normal to the deformed neutral axis and (b) the cross-sectional dimensions perpendicular to the neutral axis do not change. The original, undeformed straight beam undergoes deflections v and w. The curvatures of the projections of the deformed beam in the xy-plane and xz-plane are given by

$$\kappa_y \equiv \frac{1}{R_y} = \frac{v''}{[1+(v')^2]^{3/2}}, \quad \kappa_z \equiv \frac{1}{R_z} = \frac{w''}{[1+(w')^2]^{3/2}}, \tag{5.1}$$

where R_y and R_z are radii of curvatures and $v' = dv/dx$, and so on. We assume the deflections are small compared to the cross-sectional dimensions of the beam and slopes of the deflections are small compared to unity. These assumptions allow us to drop the quadratic terms in the curvature expressions, to have

$$\kappa_y = v'', \quad \kappa_z = w''. \tag{5.2}$$

First, let us consider the deflection in the y-direction.

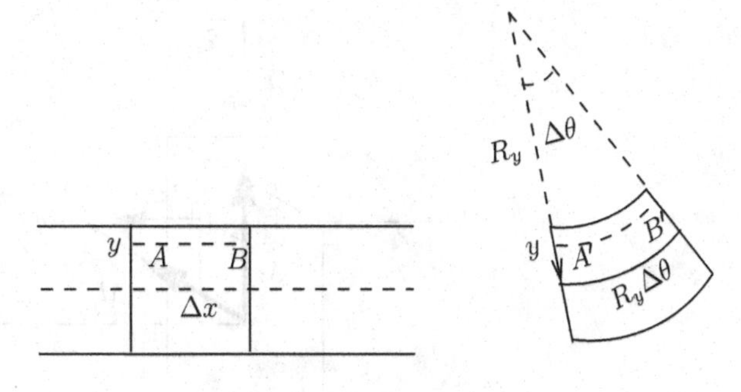

Figure 5.2 Strain due to bending.

As shown in Fig. 5.2, an element of length Δx on the neutral axis deforms into a curve of radius R_y without any change in length. Thus,

$$R_y \Delta\theta = \Delta x.$$

An element AB at a height y from the neutral axis deforms into $A'B'$ with length $(R_y - y)\Delta\theta$. Then the strain of this element is

$$\epsilon = \frac{(R_y - y)\Delta\theta - \Delta x}{\Delta x} = \frac{(R_y - y)\Delta\theta - R_y\Delta\theta}{R_y\Delta\theta}. \tag{5.3}$$

This gives

$$\epsilon = -\frac{y}{R_y} = -\kappa_y y. \tag{5.4}$$

A similar contribution to the strain can be found from the curvature in the xz-plane. The total strain is obtained as

$$\epsilon = -(\kappa_y y + \kappa_z z). \tag{5.5}$$

The axial stress in the beam is

$$\sigma = E\epsilon = -E(\kappa_y y + \kappa_z z). \tag{5.6}$$

As there is no axial force,

$$\int_A \sigma dA = 0, \tag{5.7}$$

independent of the curvatures. Here, A is the cross-sectional area of the beam. This equation determines the properties of the neutral axis:

$$\int_A E y dA = 0, \quad \int_A E z dA = 0. \tag{5.8}$$

Thus, y and z are measured from the modulus-weighted centroid of the cross-section. If E is a constant, the origin of y and z is the centroid. Keeping composite beams in mind, we keep E as a variable. From Fig. 5.3, the bending moments M_y and M_z are given by

$$M_y = \int_A \sigma z dA, \quad M_z = -\int_A \sigma y dA. \tag{5.9}$$

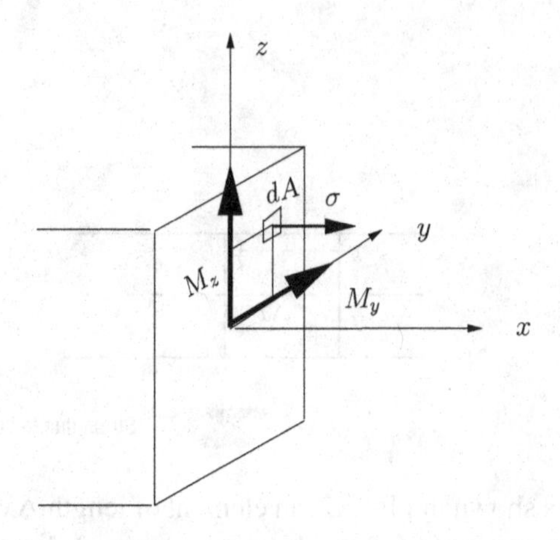

Figure 5.3 Moment components and normal stress.

Using Eq. (5.6), we obtain the moment-curvature relations

$$M_y = -\overline{EI}_{zy}\kappa_y - \overline{EI}_{yy}\kappa_z, \quad M_z = \overline{EI}_{zz}\kappa_y + \overline{EI}_{yz}\kappa_z, \tag{5.10}$$

where we have introduced the modulus-weighted second moment of areas,

$$\overline{EI}_{yy} = \int_A Ez^2 dA, \quad \overline{EI}_{zz} = \int_A Ey^2 dA, \quad \overline{EI}_{yz} = \overline{EI}_{zy} = \int_A Eyz dA. \tag{5.11}$$

We can solve for κ_y and κ_z in terms of the bending moment components, and substitute the values for κ_y and κ_z into Eq. (5.6), to find the stress as a function of applied moment components.

If there is an axis of symmetry for the cross-section (including the modulus) $\overline{EI}_{zy} = 0$, then an applied moment M_y produces a curvature in the xz-plane, κ_z, only. And, similarly, M_z produces κ_y. This way, the moment-curvature relations will be uncoupled. For the sake of simplicity, let us assume the beam is homogeneous, that is, E is a constant. Then, the neutral axis passes through the centroid of the section and the moment-curvature relations become

$$M_y = -EI_{zy}\kappa_y - EI_{yy}\kappa_z, \quad M_z = EI_{zz}\kappa_y + EI_{yz}\kappa_z. \tag{5.12}$$

Introducing the notation

$$\bar{I}^2 = I_{yy}I_{zz} - I_{yz}^2, \tag{5.13}$$

we may solve for the curvatures, to obtain

$$\kappa_y = \frac{1}{E\bar{I}^2}[I_{yy}M_z - I_{yz}M_y], \tag{5.14}$$

$$\kappa_z = -\frac{1}{E\bar{I}^2}[I_{zz}M_y + I_{yz}M_z]. \tag{5.15}$$

In the case of symmetry $I_{yz} = 0$ and we obtain the uncoupled relations

$$\kappa_y = \frac{M_z}{E I_{zz}}, \tag{5.16}$$

$$\kappa_z = -\frac{M_y}{E I_{yy}}. \tag{5.17}$$

5.0.1 Principal Axes of a Section

The coordinate system can be rotated about the x-axis, to obtain a new system: $x' = x$,

$$y' = y\cos\theta + z\sin\theta, \quad z' = -y\sin\theta + z\cos\theta. \tag{5.18}$$

In this new system, the second moments of the area are given by

$$I'_{yy} = \int_A (z')^2 dA = \int_A [s^2y^2 - 2csyz + c^2z^2]dA = I_{zz}s^2 - 2I_{yz}cs + I_{yy}c^2,$$

$$I'_{yz} = \int_A y'z'dA = \int_A [cs(z^2 - y^2) + (c^2 - s^2)yz]dA = (I_{yy} - I_{zz})cs + I_{yz}(c^2 - s^2),$$

$$I'_{zz} = \int_A (z')^2 dA = \int_A [c^2y^2 + 2csyz + z^2s^2]dA = I_{zz}c^2 + 2I_{yz}cs + I_{yy}s^2,$$

where $c = \cos\theta$ and $s = \sin\theta$. If we interchange y and z, these transformation relations are similar to the stress transformation relations involving σ_{yy}, σ_{zz}, and σ_{yz}. We choose θ to have $I'_{yz} = 0$, in order to find the principal axes of a cross-section. This leads to the equation,

$$\tan 2\theta = \frac{2I_{yz}}{I_{zz} - I_{yy}}, \tag{5.19}$$

for θ.

In the principal coordinate system

$$I'_{yy} = \frac{I_{zz} + I_{yy}}{2} + \frac{I_{yy} - I_{zz}}{2}\cos 2\theta - I_{yz}\sin 2\theta,$$

$$I'_{zz} = \frac{I_{zz} + I_{yy}}{2} - \frac{I_{yy} - I_{zz}}{2}\cos 2\theta + I_{yz}\sin 2\theta.$$

In the new coordinate system the moment-curvature relations will be uncoupled. Of course, we also have to to find the components of the moment and the curvature in the new system. Thus,

$$M'_y = -EI'_{yy}\kappa'_z, \quad M'_z = EI'_{zz}\kappa'_y. \tag{5.20}$$

5.0.2 Example: Angle Section

For the angle section shown in Fig. 5.4 the Young's modulus is homogeneous. A moment component $M_z = M_0$ is applied at this section. First obtain the location of the neutral axis, then find the second moments and the product moment, and finally compute the stresses at the corner points A, B, and C. Also, find the

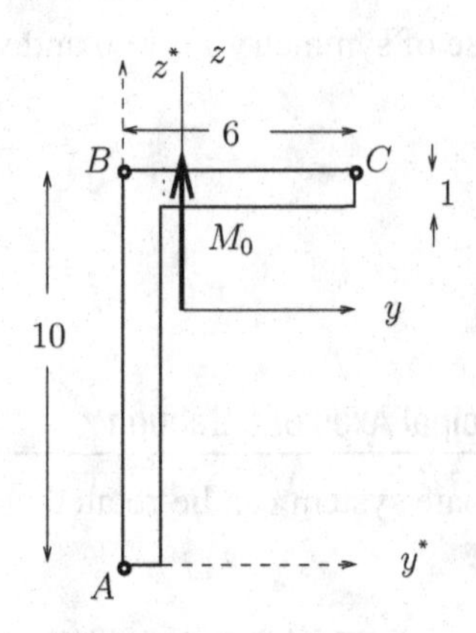

Figure 5.4 Angle section beam.

principal axes for this section and using them obtain the stresses at the corner points. All dimensions are in centimeters. When the Young's modulus is constant, the neutral axis is through the centroid. To find the centroid, we use a temporary coordinate system with y^*, z^* as shown in Fig. 5.4. We divide the section into two areas: $A_1 = 10 \times 1$ and $A_2 = 5 \times 1$. The total area is

$$A = A_1 + A_2 = 15 \text{ cm}^2.$$

Taking the moment of the areas about the z'-axis:

$$A\bar{y}^* = 10 \times 0.5 + 5 \times 3.5, \quad \bar{y}^* = 1.5.$$

$$A\bar{z}^* = 10 \times 5 + 5 \times 9.5, \quad \bar{z}^* = 6.5.$$

With these coordinates, we establish our coordinate system y, z at the centroid. The coordinates of the centroids of the two areas are obtained as $(-1, -1.5)$ and $(2, 3)$. The second moments are found using the shift theorem (see your statics text if you have forgotten this) as

$$I_{yy} = \frac{1 \times 10^3}{12} + 10 \times 1.5^2 + \frac{5 \times 1^3}{12} + 5 \times 3^2 = 151.25,$$

$$I_{zz} = \frac{10 \times 1^3}{12} + 10 \times 1^2 + \frac{1 \times 5^3}{12} + 5 \times 2^2 = 41.25,$$

$$I_{yz} = 10 \times (-1)(-1.5) + 5 \times 2 \times 3 = 45.$$

From the moment-curvature relations (5.10), we have

$$I_{yz}\kappa_y + I_{yy}\kappa_z = 0, \quad I_{zz}\kappa_y + I_{yz}\kappa_z = M_0/E.$$

Using the numerical values for the second moments, these become

$$45\kappa_y + 151.25\kappa_z = 0, \quad 41.25\kappa_y + 45\kappa_z = M_0/E.$$

Solving these equations, we get

$$\kappa_y = 0.03589 M_0/E, \quad \kappa_z = -0.01068 M_0/E.$$

The bending stress distribution across this section is given by Eq. (5.6) as

$$\sigma = -M_0(0.03589y - 0.01068z).$$

To compute the stresses at the points A, B, and C, we note their coordinates as

$$A : (-1.5, -6.5), \quad B : (-1.5, 3.5), \quad C : (4.5, 3.5),$$

to obtain

$$\sigma_A = -0.0156M_0, \quad \sigma_B = 0.0912M_0, \quad \sigma_C = -0.1241M_0.$$

We may also solve this problem using the principal coordinates. The needed rotation angle is obtained from

$$\tan 2\theta = \frac{2I_{yz}}{I_{zz} - I_{yy}} = \frac{90}{41.25 - 151.25} = -0.8182,$$

which gives $2\theta = -0.6857$ or $\theta = -0.3429$ radians $= -19.64°$. There is a second solution that is 90° away from this. The second solution just rotates the coordinate orientations obtained using the first solution by 90°. Let y' and z' represent the rotated coordinates. Of course, $I'_{yz} = 0$ and

$$I'_{yy} = \frac{151.25 + 41.25}{2} + \frac{151.25 - 41.25}{2}\cos 2\theta - 45\sin 2\theta = 167.31,$$

$$I'_{zz} = \frac{151.25 + 41.25}{2} - \frac{151.25 - 41.25}{2}\cos 2\theta + 45\sin 2\theta = 25.19,$$

The components of the applied moment along the principal coordinates are

$$M'_y = M_0 \sin\theta = -0.3362M_0, \quad M'_z = M_0\cos\theta = 0.9418M_0.$$

Curvatures are given by

$$\kappa'_y = \frac{M'_z}{EI'_{zz}} = 0.9418M_0/(25.19E) = 0.0374M_0/E,$$

$$\kappa'_z = -\frac{M'_y}{EI'_{yy}} = 0.3369M_0/(167.31E) = 0.0020M_0/E.$$

The stress distribution is given by

$$\sigma = -E(\kappa'_y y' + \kappa'_z z') = -M_0(0.0374y' + 0.0020z').$$

The new coordinates of the corner points are

$$A : (-1.5\cos\theta - 6.5\sin\theta,\ 1.5\sin\theta - 6.5\cos\theta) = (0.7725, -6.6260),$$

$$B : (-2.5894, 2.7920), \quad C : (3.0614, 4.8091).$$

With these coordinates, we obtain the same stress values at the corner points. As can be seen, the method of principal axes involves considerably more calculations as the applied moment as well as the coordinates of the points have to be resolved along the principal axes.

5.1 Equilibrium Equations

As shown in Fig. 5.5, the applied distributed loads, p_y along the y-direction and p_z along the z-direction, create shear forces V_y and V_z and bending moments M_y and M_z.

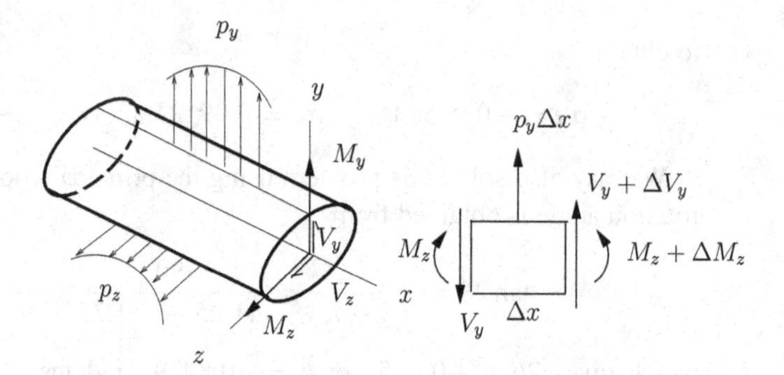

Figure 5.5 Force distribution and the equilibrium of an element.

Force and moment balances of an element of the beam of length Δx give the equilibrium equations

$$\frac{dV_y}{dx} + p_y = 0, \quad \frac{dV_z}{dx} + p_z = 0, \tag{5.21}$$

$$\frac{dM_z}{dx} + V_y = 0, \quad \frac{dM_y}{dx} - V_z = 0. \tag{5.22}$$

The negative sign in the last equation can be understood if we rotate the figure to have p_z pointing upward. This rotation results in M_y pointing away from the page, unlike M_z.

In integrating these equations to find the bending moment, the boundary conditions on the beam have to be considered. A given end of the beam can have (a) prescribed shear force and prescribed moment, (b) prescribed vertical displacement and prescribed slope, (c) prescribed shear force and prescribed slope, and (d) prescribed bending moment and prescribed displacement.

If the shear force and bending moment distributions can be found using only the equilibrium equations, we have a statically determinate problem. If the moment curvature relations

$$EI_{yy}w'' = -M_y, \quad EI_{zz}v'' = M_z, \tag{5.23}$$

have to be integrated to find the shear force and bending moment, we have a statically indeterminate problem.

Let us examine some examples of shear force and bending moment calculations for statically determinate problems first. For simplicity we assume symmetry about the z-axis, and use the notations: $I = I_{zz}$, $V = V_y$, and $M = M_z$.

5.1.1 Example: Elliptic Lift Distribution

We assume a uniform wing is loaded by a lift distribution that has the shape of an ellipse,

$$p(x) = p_0\sqrt{1-(x/L)^2},$$

where p_0 is the maximum load density at $x = 0$. This is shown in Fig. 5.6. The shear force distribution is found from

$$\frac{dV}{dx} = -p_0[1-(x/L)^2]^{1/2}.$$

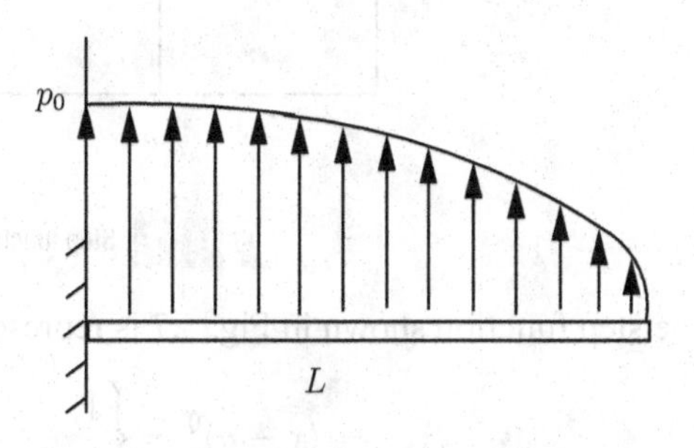

Figure 5.6 Elliptic loading on a wing.

A change of variable

$$\frac{x}{L} = \sin\theta, \quad dx = L\cos\theta d\theta,$$

facilitates the integration of the load distribution.

$$V = -p_0L\int_{\pi/2}^{\theta}\cos^2\theta d\theta = -p_0L\int_{\pi/2}^{\theta}\frac{1+\cos 2\theta}{2}d\theta,$$

$$= -p_0L\left[\frac{\theta-\pi/2}{2}+\frac{\sin 2\theta}{4}\right] = -\frac{p_0L}{4}[2\theta-\pi+\sin 2\theta].$$

The differential equation for the bending moment,

$$\frac{dM}{dx} = -V,$$

gives, after integration by parts,

$$M = \frac{p_0L^2}{4}\int_{\pi/2}^{\theta}[2\theta-\pi+\sin 2\theta]\cos\theta d\theta,$$

$$= \frac{p_0L^2}{4}\left[(2\theta-\pi)\sin\theta+2\cos\theta-\frac{2}{3}\cos^3\theta\right].$$

The maximum values of the shear force and the bending moment are at $x = 0$, given by

$$V_{max} = \frac{\pi p_0L}{4}, \quad M_{max} = \frac{p_0L^2}{3}.$$

5.1.2 Concentrated Forces and Moments

In many beam problems the applied distributed force density $p(x)$ consists of functions with multiple discontinuities in the form of steps or it is idealized as a concentrated force or concentrated moment. The Macaulay bracket notation is a simple way to represent these situations.

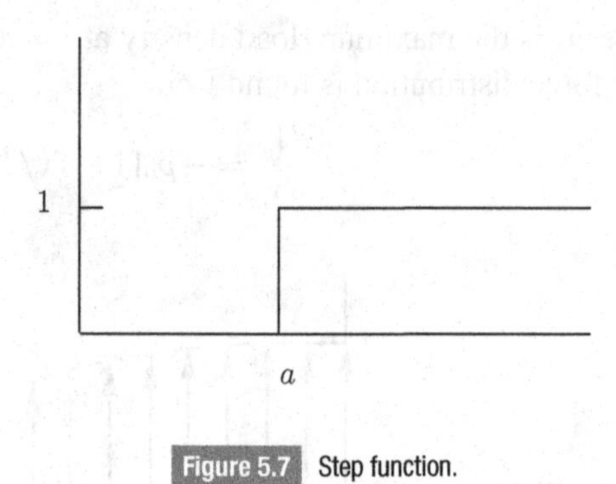

Figure 5.7 Step function.

The step function shown in Fig. 5.7 is represented as

$$\langle x-a\rangle^0 = \begin{cases} 1, & x > a, \\ 0 & x < a. \end{cases}$$

The integral of this function is, of course, zero up to any $x < a$ and it is linearly increasing with slope of unit for $x > a$. We write this as

$$\langle x-a\rangle^1 = \int^x \langle x-a\rangle^0 dx = \begin{cases} x-a, & x > a, \\ 0 & x < a. \end{cases} \tag{5.24}$$

Continuing this process, we have

$$\begin{aligned} \frac{1}{n+1}\langle x-a\rangle^{n+1} &= \int^x \langle x-a\rangle^n dx \\ &= \begin{cases} \dfrac{(x-a)^{n+1}}{n+1}, & x > a, \\ 0 & x < a. \end{cases}, \quad n \geq 0. \end{aligned} \tag{5.25}$$

We know concentrated forces cause a jump in the shear force diagram. To represent them as a distributed force we have to use a generalized function, which is the derivative of the step function. Using the Macaulay bracket notation we write this as

$$\langle x-a\rangle^{-1} = \frac{d}{dx}\langle x-a\rangle^0, \tag{5.26}$$

which is also called a singularity function or Dirac delta function as it goes to infinity when $x = a$ and is zero everywhere else. We may go one step further and also include concentrated moments using the "derivative" of the Dirac delta function,

$$\langle x-a\rangle^{-2} = \frac{d}{dx}\langle x-a\rangle^{-1}. \tag{5.27}$$

To emphasize the fact that this is just a notation, we do not use a minus sign in front of the derivative.

5.1.3 Example: Concentrated Force and Moment

Consider the loaded beam shown in Fig. 5.8(a).

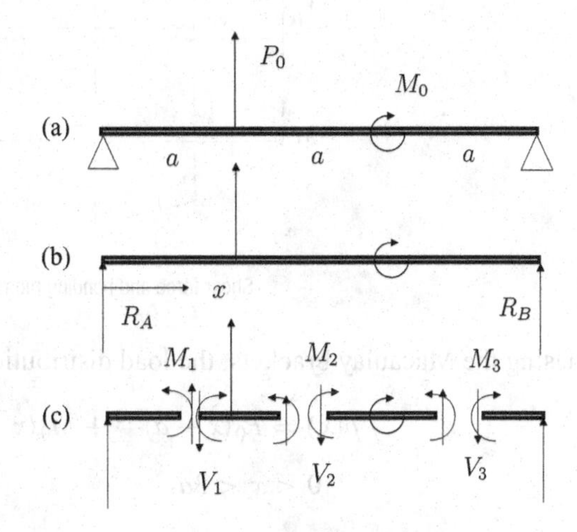

Figure 5.8 Simply supported beam loaded with a concentrated force and a moment with free-body diagrams to calculate shear forces and bending moments.

First, using Fig. 5.8(b), let us obtain the shear force and the bending moment distributions using free-body diagrams. Let R_A and R_B be the reactions from the supports. Setting the moment about A and about B, respectively, to zero,

$$R_B = \frac{M_0}{3a} - \frac{P_0}{3}, \quad R_A = -\frac{M_0}{3a} - \frac{2P_0}{3}.$$

The three cuts shown in Fig. 5.8(c) have to be considered sequentially. From the first cut,

$$V_1 = \frac{M_0}{3a} + \frac{2P_0}{3}, \quad M_1 = -\left[\frac{M_0}{3a} + \frac{2P_0}{3}\right] x.$$

The other two cuts give

$$V_2 = \frac{M_0}{3a} - \frac{P_0}{3}, \quad M_2 = -\left[\frac{M_0}{3a} + \frac{2P_0}{3}\right] x + P_0(x - a),$$

$$V_3 = \frac{M_0}{3a} - \frac{P_0}{3}, \quad M_3 = \left[\frac{M_0}{3a} - \frac{P_0}{3}\right] (3a - x).$$

We note that

$$V_2 - V_1 = -P_0,$$

and at $x = 2a$

$$M_3 - M_2 = M_0.$$

The shear force and the bending moment diagrams are shown in Fig. 5.9. We have assumed $M_0 > P_0 a$ for drawing the diagrams.

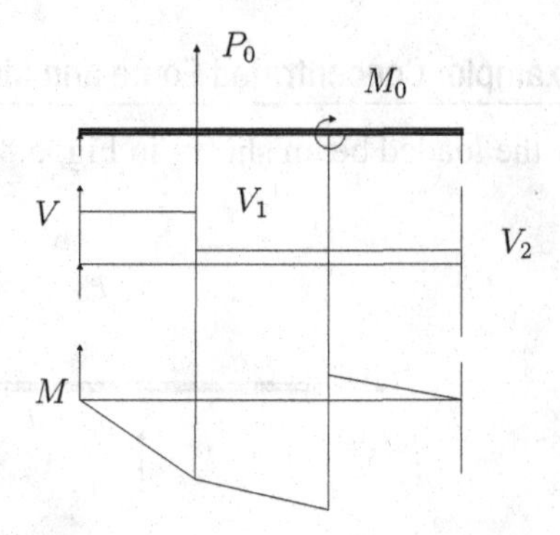

Figure 5.9 Shear force and bending moment diagrams.

Using the Macaulay brackets, the load distribution $p(x)$ can be represented as

$$p(x) = P_0\langle x - a\rangle^{-1} + M_0\langle x - 2a\rangle^{-2}.$$
$$0 < x < 3a. \tag{5.28}$$

Note that we have excluded the reactions at the supports. Integrating $p(x)$, we get

$$-V(x) = P_0\langle x - a\rangle^{0} + M_0\langle x - 2a\rangle^{-1} + C_1, \tag{5.29}$$

where C_1 is a constant of integration. Integrating again,

$$M = P_0\langle x - a\rangle^{1} + M_0\langle x - 2a\rangle^{0} + C_1 x + C_2, \tag{5.30}$$

where C_2 is a second constant. For the simply supported beam, the boundary conditions are: $M(0) = 0$ and $M(3a) = 0$. The first of these conditions gives

$$C_2 = 0. \tag{5.31}$$

The second condition gives

$$P_0(2a) + M_0 + 3aC_1 = 0, \quad \text{or} \quad C_1 = -(2aP_0 + M_0)/(3a). \tag{5.32}$$

Then

$$-V = P_0\langle x - a\rangle^{0} + M_0\langle x - 2a\rangle^{-1} - \frac{2P_0}{3} - \frac{M_0}{3a},$$
$$M = P_0\langle x - a\rangle^{1} + M_0\langle x - 2a\rangle^{0} - \left[\frac{2P_0}{3} + \frac{M_0}{3a}\right]x.$$

A comparison of these expressions with the ones obtained using the free-body diagrams shows they are identical. Thus, the Macaulay bracket notation allows us to skip the free-body diagrams and to write the load $p(x)$ in terms of singularity functions, which can be integrated to get the shear force and the bending moment.

5.2 Shear Stresses in Beams

The calculation of shear stresses due to bending in an arbitrary cross-section beam is a complicated problem in elasticity and there are no general formulas. In special cases such as a rectangular cross-section we have the parabolic distribution of shear stresses. Here we consider thin-walled open- and closed-cell tubes. Consider a section of the beam of length Δx as shown in Fig. 5.10. Under the thin-wall

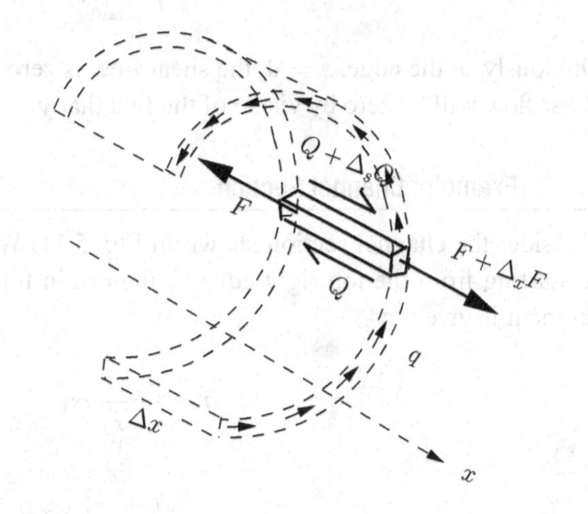

Figure 5.10 Shear flow and bending stress balance in a beam element.

assumption we approximate the shear stress distribution across the thickness t is uniform.

5.2.1 Open-cell Tubes

Then,

$$q = \tau t. \tag{5.33}$$

An element taken at s, with width Δs, has normal force F on one side and $F + \Delta_x F$ on the other. The shear flow creates a force Q at the lower edge and $Q + \Delta_s Q$ at the upper edge. We have

$$F = \sigma t \Delta s, \quad Q = q \Delta x. \tag{5.34}$$

The force balance relation,

$$\Delta_x F + \Delta_s Q = 0, \tag{5.35}$$

can be written as

$$\Delta_x \sigma t \Delta s + \Delta_s q \Delta x = 0 \quad \text{or} \quad \frac{\partial \sigma}{\partial x} t + \frac{\partial q}{\partial s} = 0. \tag{5.36}$$

At the two edges of the open cell there are no shear stresses and $q = 0$ at these edges. Using this, we can integrate this equation along s to find q. For example, for an open section that is symmetric about the z-axis, using $M = M_{zz}$, $I = I_{zz}$,

$V = V_y$, we have

$$\sigma = -\frac{M}{I}y, \quad \frac{\partial \sigma}{\partial x} = -\frac{\partial M}{\partial x}\frac{y}{I} = \frac{Vy}{I}. \tag{5.37}$$

and

$$q = -\frac{V}{I}\int_{s=0}^{s} ytds. \tag{5.38}$$

Obviously, at the edge, $s = 0$, the shear flow is zero. At the other edge, $s = s_{max}$, shear flow will be zero by virtue of the fact that y is measured from the centroid.

5.2.2 Example: Channel Section

Consider the channel section shown in Fig. 5.11. We assume uniform thickness t. Starting from the top right edge as the origin for s, the shear flow in the top segment is given by

$$q = -\frac{Vt}{I}as,$$

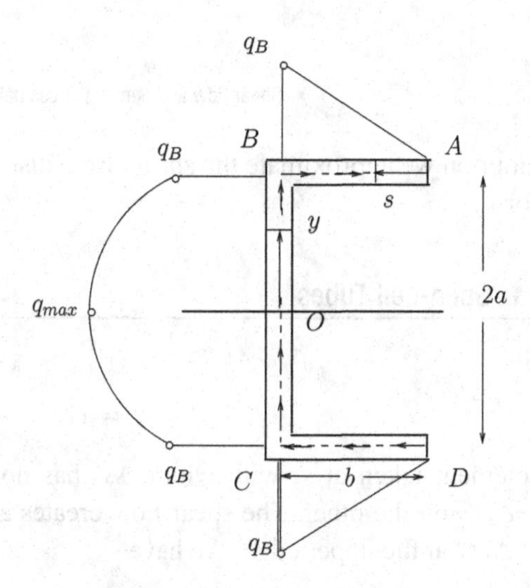

Figure 5.11 Shear flow in a channel section.

where I is given by

$$I = \frac{2}{3}a^3t + 2a^2bt = \frac{2}{3}a^3t\left[1 + \frac{3b}{a}\right].$$

Here, the thin wall assumption allows us to neglect the t^3-terms. At the top corner B, where $s = b$,

$$q_B = -\frac{Vt}{I}ab.$$

Along the vertical segment, it is convenient to use a new coordinate y with B being at $y = a$. Then in BC,

$$q = q_B - \frac{Vt}{I}\int_a^y y(-dy) = -\frac{Vt}{I}\left[ab - \frac{y^2 - a^2}{2}\right].$$

The magnitude of the shear flow, which varies as a parabola in BC, is a maximum at $y = 0$.

$$q_{max} = -\frac{Vt}{I}\left[ab + \frac{a^2}{2}\right].$$

When $y = -a$, we have

$$q_C = q_B.$$

Further, along the lower segment from C to D it decreases to be zero at the free edge, D. A plot of the variation of the shear flow is superposed in Fig. 5.11. Along the mid-surface the shear flow direction is indicated using arrows. As s is taken counterclock-wise the negative values of q show that the shear flow is clockwise. The net force from the horizontal segments cancel each other out, resulting zero horizontal force. The vertical force is given by

$$\int(-q)dy = \int_{y=-a}^{y=a} \frac{Vt}{I}\left[ab - \frac{y^2 - a^2}{2}\right] dy = \frac{2Vt}{I}\left[a^2 b - \frac{a^3}{3} - a^3\right] = V.$$

This verifies the fact that the applied shear force is the resultant of the distributed shear flow.

5.2.3 Center of Shear

In the example of a channel section, we saw that the shear flow distribution creates a vertical force equal to the applied shear force. A question remains: Where is the location along the z-axis where the shear force V has to be applied to create the same moment as that due to the shear flow distribution? This location is called the shear center. Remember the normal stress and the shear flow distribution are derived on the basis that the beam bends in the y-direction without any rotation. We may compute the moment produced by the shear flow distribution about the point O. The force due to the shear flow in the upper segment is given by the area of the triangle in Fig. 5.11,

$$F = \frac{Vt}{2I}ab^2.$$

There is an equal and opposite force due to the shear flow in the lower segment. Then the torque is given by $T = 2aF$, or

$$T = \frac{Vt}{I}a^2b^2.$$

To create the same torque the shear force has to be applied at a distance e to the left of O, such that

$$Ve = T.$$

From this the distance to the shear center S is given by

$$e = \frac{a^2 b^2 t}{I} = \frac{3b^2}{2a\,[1 + 3b/a]} = \frac{b}{2[1 + a/(3b)]}.$$

This expression shows that $e = 0$ when $b = 0$ and $e = b/2$ when b is very large compared to a.

In the general case of a cross-section with symmetry about the z-axis, we compute the torque produced by the shear flow distribution about a convenient point O along the z-axis and find the distance to the shear center e from

$$Ve = -\int qpds, \tag{5.39}$$

where p is the perpendicular distance to the tangent at a point s, and e is measured along the positive z-direction.

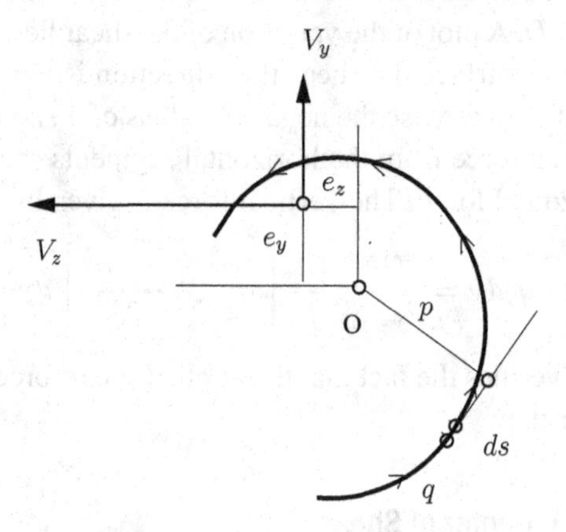

Figure 5.12 Shear flow and the shear center location.

For sections without any symmetry axis, we can apply arbitrary shear forces V_y and V_z to compute the shear flow q. Then the torque due to q is found about any point O to have

$$V_y e_z - V_z e_y = -\int qpds, \tag{5.40}$$

where e_y and e_z are the coordinates of the shear center. Fig. 5.12 shows a schematic of this case.

5.2.4 Example: Split Circular Beam

Consider a circle of radius R and thickness t with small gap at $\theta = 0$. This is shown in Fig. 5.13. For a point on the wall with angle θ,

$$y = R \sin\theta. \tag{5.41}$$

The second moment of the area about the horizontal axis is

$$I = \int_0^{2\pi} y^2 t R d\theta = R^3 t \int_0^{2\pi} \sin^2\theta d\theta = \pi R^3 t. \tag{5.42}$$

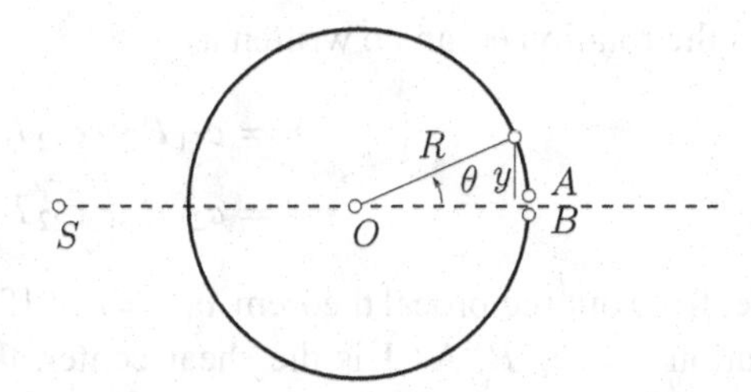

Figure 5.13 A split circular beam subjected to a vertical shear force.

Starting from the free edge A, the shear flow at θ is obtained as

$$q = -\frac{Vt}{I}\int_0^{\theta} yR d\theta = -\frac{VR^2 t}{I}\int_0^{\theta} \sin\theta d\theta = \frac{VR^2 t}{I}[\cos\theta - 1]. \quad (5.43)$$

Taking counterclockwise moments about the center O, the torque due to q is

$$T = \int_0^{2\pi} qR^2 d\theta = \frac{VR^4 t}{I}\int_0^{2\pi} [\cos\theta - 1] d\theta = -\frac{2\pi VR^4 t}{I} = -2R. \quad (5.44)$$

To create an equal torque the applied shear force should act through the shear center S, which is at a distance $2R$, left of the center O.

In many cases, we see the shear center is located "outside" the cross-section. If a cantilever beam with a tip load V has this split circular cross-section, the loading has to be effected through a rigid attachment welded perpendicular to the beam.

5.2.5 Center of Shear and the Center of Twist

In Chapter 4 we encountered the center of twist as the point about which a thin-walled cross-section rotates under a torque. The center of shear is the point through which shear force has to be applied so that the beam bends without any rotation. We can show that these two points coincide, by using the energy principles.

We begin by assuming the shear center S and the center of twist C are two distinct points, say point 1 and point 2, as shown in Fig. 5.14. Next we apply a vertical load P at 1 and a torque T at 2. The vertical deflection v under the load

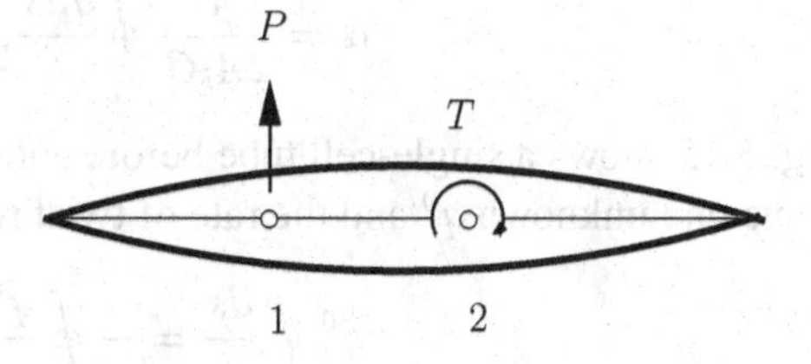

Figure 5.14 Center of shear 1 and center of twist 2.

P and the rotation θ can be written as

$$v = c_{11}P + c_{12}T, \tag{5.45}$$

$$\theta = c_{21}P + c_{22}T, \tag{5.46}$$

where, from our reciprocal theorem, $c_{12} = c_{21}$. If we have P alone, it will produce a rotation $\theta = c_{21}P$. As 1 is the shear center, the beam should deflect without rotation. Then $\theta = 0$. This implies $c_{12} = 0$.

Now, let us try applying T alone. As the vertical deflection, v, due to T is given by $v = c_{12}T$, we will have $v = 0$. But the beam is rotating about the point 2. In order to have no deflection at 1, the points 1 and 2 must be the same.

Considering the aircraft wing as a thin-walled beam, we may analyze two scenarios: (a) the lift load passes through a point in front of the shear center and (b) it acts through a point behind the shear center. In the first case if the lift increases by a small amount it produces a torque and the angle of attack would also increase. This results in more lift. This is a destabilizing process, straining the wing to the point of failure. In the second case, any increase in lift would reduce the angle of attack and decrease the lift, which is a stabilizing process. This example shows the importance of determining the shear center accurately during the design process so that the lift acts behind the shear center.

5.3 Multi-cell Tubes

Unlike an open-cell tube, closed-cell tubes have no stress-free edges, $s = 0$, to begin the shear flow evaluation using the formula

$$q = -\frac{V}{I}\int_{s=0}^{s} yt\,ds. \tag{5.47}$$

In such cases, we proceed in two steps: (a) we make a sufficient number of cuts at convenient points (the cut cells must have certain symmetry if possible) to transform all cells to open cells and compute the varying shear flow distribution, q^v, using Eq. (5.47) with $q^v = 0$ at the newly created stress-free edges; and (b) apply unknown constant shear flows, q_i^0, around the ith cell. In fact, these are the shear flows at the cuts. Finally, we solve for the unknown shear flows, q_i^0, using the requirement that there is no rotation during bending with the shear force acting through the shear center. Recall that the rate of twist of the ith cell is given by

$$\alpha = \frac{1}{2A_i G}\oint_i \frac{q\,ds}{t}. \tag{5.48}$$

Fig. 5.15 shows a single-cell tube before and after making a cut. With one cut we have one unknown q^0 and the rate of twist relation becomes

$$q^0 \oint \frac{ds}{t} = -\oint \frac{q^v ds}{t}, \tag{5.49}$$

where q^v is given by Eq. (5.47).

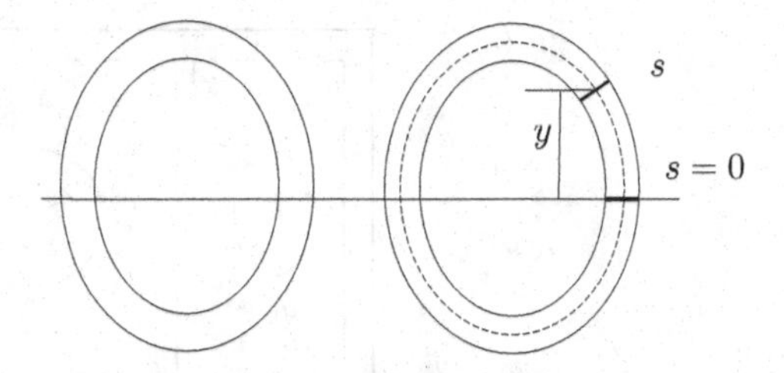

Figure 5.15 A single-cell tube with and without a cut.

5.3.1 Shear Flow Transfer at Wall Intersections

When three or more walls meet at a node, as shown in Fig. 5.16, a force balance of the node in the axial direction establishes a relation among the shear flows at the node. For example, at node 2,

$$q_{23}^{v} = q_{12}^{v} + q_{82}^{v} + q_{92}^{v}. \tag{5.50}$$

The constant shear flows q_i^0 are handled in the same way as we did in torsion. We use the electrical current analogy and obtain a shear flow of $q_1^0 - q_2^0$ in the wall between cell 1 and cell 2. This is also shown in Fig. 5.16.

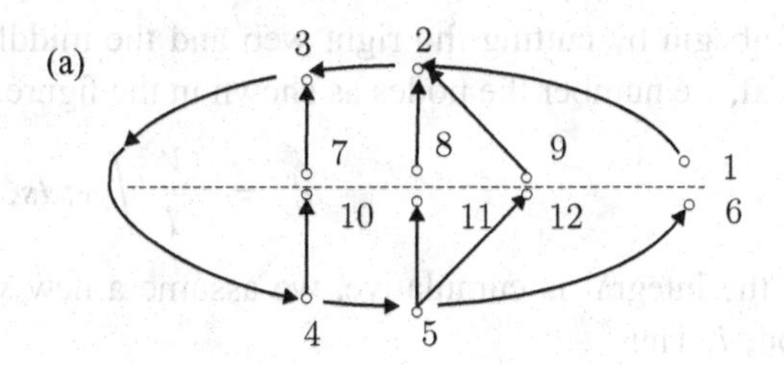

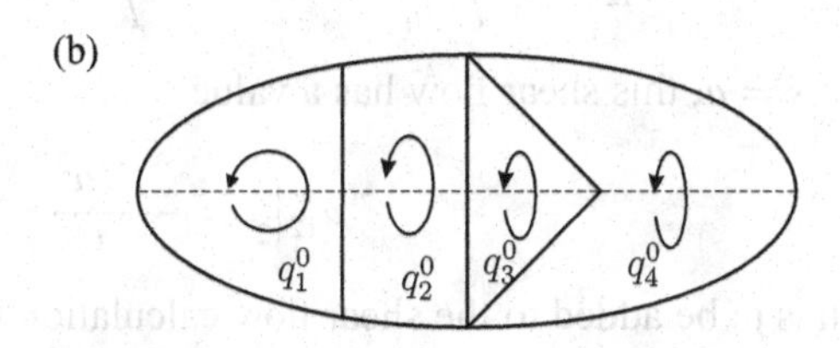

Figure 5.16 A four-cell tube with (a) and without (b) cuts.

5.3.2 Example: Two-cell Tube

Obtain the shear flows and the shear center for the two-cell tube shown in Fig. 5.17. Assume a thickness of t for the middle web and $2t$ for the remaining walls. The smaller cell is $2a \times a$ and the larger cell is $2a \times 2a$ in dimension.

The value of I under the thin-wall assumption is obtained as

$$I = 2\frac{2a^3 \times 2t}{3} + \frac{2a^3 \times t}{3} + 2 \times 2ta \times a^2 + 2 \times 4ta \times a^2 = \frac{46}{3}a^3 t.$$

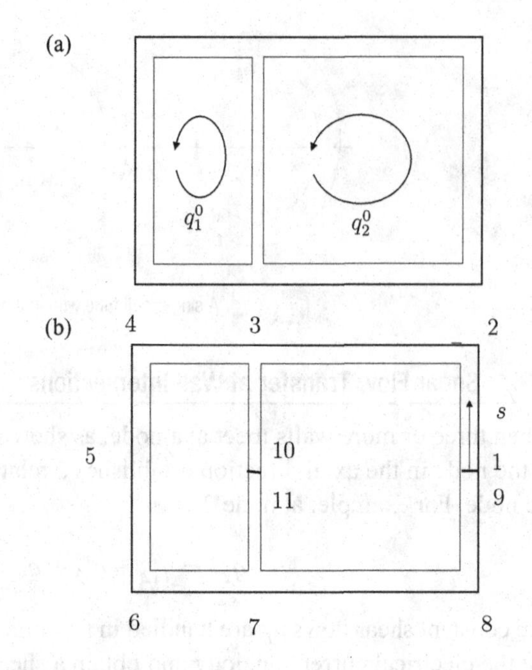

Figure 5.17 A two-cell tube without (a) and with (b) cuts.

We begin by cutting the right web and the middle web to create an open cell. Next, we number the nodes as shown in the figure. In computing

$$q_{ij}^v = -\frac{V}{I}\int ytds,$$

as the integral is cumulative, we assume a new s coordinate starting from the node i. Thus,

$$q_{12}^v = -\frac{V}{I}\int_0^s y2tds = -\frac{Vt}{I}\int_0^s 2sds = -\frac{Vt}{I}s^2.$$

When $s = a$, this shear flow has a value

$$q_{12}^v\big|_2 = -\frac{Vta^2}{I},$$

which is to be added to the shear flow calculation from 2 to 3.

$$q_{23}^v = -\frac{Vt}{I}\left[a^2 + \int_0^s 2ads\right] = -\frac{Vt}{I}[a^2 + 2as], \quad q_{23}^v\big|_3 = -\frac{5Vta^2}{I}.$$

$$q_{103}^v = -\frac{Vts^2}{2I}, \quad q_{103}^v\big|_3 = -\frac{Vta^2}{2I}.$$

$$q_{34}^v = -\frac{Vt}{I}\left[\frac{a^2}{2} + 5a^2 + \int a2ds\right] = -\frac{Vt}{I}\left[\frac{a^2}{2} + 5a^2 + 2as\right].$$

$$q_{34}^v\big|_4 = -\frac{15Vta^2}{2I}.$$

Along 45, the value of y is $(a-s)$ and

$$q_{45}^{v} = -\frac{Vt}{I}\left[\frac{15a^2}{2} + \int_0^s (a-s)2ds\right] = -\frac{Vt}{I}\left[\frac{15a^2}{2} + a^2 - (a-s)^2\right],$$

$$q_{45}^{v}\big|_5 = -\frac{17Vta^2}{2I}.$$

For the lower half of the beam the value of y is negative and the integrals contribute negative values to q^v. Due to this, we get the same values for q^v as on the upper half, except the direction of the shear flow is opposite what is in the upper half, that is, from left to right or from top to bottom. Fig. 5.18 shows the magnitudes (not to scale) and directions of the shear flows, q^v, after removing a factor Vta^2/I.

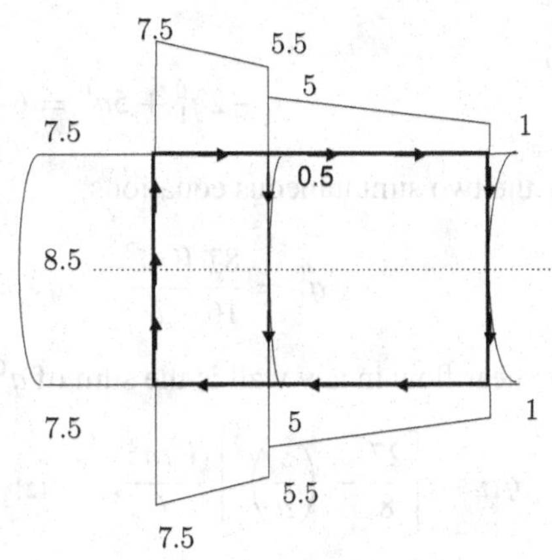

Figure 5.18 The shear flow distribution after making the cuts (the magnitudes are not to scale).

The shear forces produced by these shear flows are obtained as

$$F_{ij} = \int_i^j q^v ds,$$

which are the areas under the magnitude curves shown in Fig. 5.18. All our computed shear flows have a minus sign in front. We include this in computing the shear forces:

$$F_{21} = \frac{1}{3}\frac{Vta^3}{I}, \quad F_{32} = 6\frac{Vta^3}{I}, \quad F_{43} = \frac{13}{2}\frac{Vta^3}{I},$$

$$F_{54} = \frac{49}{6}\frac{Vta^3}{I}, \quad F_{310} = \frac{1}{6}\frac{Vta^3}{I}.$$

As a check, we may compute the net vertical force due to the shear flows.

$$2(F_{12} + F_{54} + F_{103}) = 2\left(-\frac{1}{3} + \frac{49}{6} - \frac{1}{3} - \frac{1}{6}\right)\frac{Vta^3}{I} = V.$$

Again, the contributions from the lower half have the same magnitude but the opposite sign.

The condition that the rate of twist of each of the two cells must be zero, that is,

$$\oint q ds/t = 0,$$

is used to obtain the shear flows at the cuts.

$$q_1^0\left[\frac{4a}{2t}+\frac{2a}{t}\right]-q_2^0\left[\frac{2a}{t}\right]=\frac{2F_{54}}{2t}+\frac{2F_{43}}{2t}+\frac{F_{310}}{t},$$

which can be simplified to get

$$4q_1^0-2q_2^0=15\frac{Vta^2}{I}.$$

From the second cell,

$$-q_1^0\left[\frac{2a}{t}\right]+q_2^0\left[\frac{6a}{2t}+\frac{2a}{t}\right]=\left[\frac{-2F_{310}}{t}+\frac{2F_{32}}{2t}+\frac{2F_{21}}{2t}\right],$$

or

$$-2q_1^0+5q_2^0=6\frac{Vta^2}{I}.$$

Solving the two simultaneous equations,

$$q_1^0=\frac{87}{16}\frac{Vta^2}{I},\quad q_2^0=\frac{27}{8}\frac{Vta^2}{I}.$$

The net shear flow in any wall is the sum of q^0 and q^v. Thus

$$q_{12}=\left[\frac{27}{8}-\left(\frac{s}{a}\right)^2\right]\frac{Vta^2}{I},\quad q_{12}|_2=\frac{19}{8}\frac{Vta^2}{I},$$

$$q_{23}=\left[\frac{19}{8}-2\frac{s}{a}\right]\frac{Vta^2}{I},\quad q_{23}|_3=-\frac{13}{8}\frac{Vta^2}{I},$$

$$q_{103}=\left[\frac{87}{16}-\frac{27}{8}-\frac{1}{2}\left(\frac{s}{a}\right)^2\right]\frac{Vta^2}{I},\quad q_{103}|_3=\frac{25}{16}\frac{Vta^2}{I},$$

$$q_{34}=\left[\frac{25}{16}-\frac{13}{8}-2\frac{s}{a}\right]\frac{Vta^2}{I},\quad q_{34}|_4=-\frac{33}{16}\frac{Vta^2}{I},$$

$$q_{45}=\left[-\frac{33}{16}+1-\left(\frac{s-a}{a}\right)^2\right]\frac{Vta^2}{I},\quad q_{45}|_5=-\frac{17}{16}\frac{Vta^2}{I}.$$

Having obtained all the shear flows, we are in a position to compute the shear center. For this we choose a convenient point to compute the torque produced by the shear flows. The point 5 is appropriate for this as the shear force F_{45} passes through it. The net torque is

$$T=2A_1q_1^0+2A_2q_2^0-\left[2\times\frac{13}{2}-2\times 6-3\times\frac{2}{3}-\frac{1}{3}\right]\frac{Vta^4}{I},$$

$$=\frac{257}{12}\frac{Vta^4}{I}=\frac{257}{184}Va.$$

The applied vertical force V must applied through a point, $257a/184$ to the right of the node 5.

5.4 Sheet-stringer Construction

In aircraft structures, except for the spars, solid section beams are seldom used. The bending stresses are carried by stringers (long metal bars with cross-sections in the shapes of "L," "Z," or "T"). Sometimes these are also called longerons. These are kept in position by outer metal sheets made of aluminum. In a preliminary design the exact shape of the stringer cross-section is not important but their areas and the locations of their centroids are. We use circular cross-sections, which make them look like the booms used to support sails in boats. The stress analysis of this simplified structure is called boom analysis. Our basic assumption is that the sheets carry shear stresses and the booms carry normal stresses.

As a simple example, let us consider two booms of area B connected by a sheet of thickness t and height $2c$. Under the elementary beam theory assumptions the axial strain is distributed as

$$\epsilon = -\kappa y, \tag{5.51}$$

which gives the stress at the centroid of the boom 1,

$$\sigma = -\kappa c, \tag{5.52}$$

and at the centroid of boom 2,

$$\sigma = \kappa c. \tag{5.53}$$

The axial force in the top boom is

$$F = \sigma B. \tag{5.54}$$

The applied moment M is given by

$$M = -2cF = -2c\sigma B = 2c^2 B\kappa. \tag{5.55}$$

We may set

$$I = 2Bc^2, \tag{5.56}$$

to have

$$\kappa = \frac{M}{I}, \quad \sigma = -\frac{Mc}{I}, \quad F = -\frac{MBc}{I}. \tag{5.57}$$

As shown in Fig. 5.19, for an infinitesimal length dx, the change in force

$$dF = d\sigma B = -dMBc/I = qdx, \tag{5.58}$$

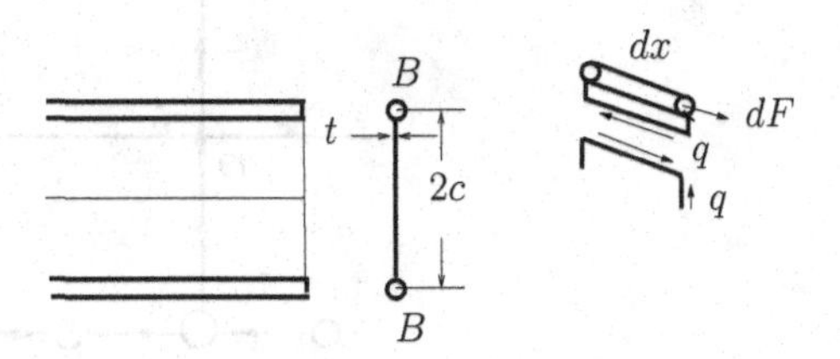

Figure 5.19 Two booms connected by a web with the shear flow produced by a varying axial stress.

and

$$q = -\frac{dM}{dx}\frac{Bc}{I} = -\frac{VBc}{I}. \tag{5.59}$$

The shear flow is a constant between the two booms.

5.4.1 Example: Open Cell

Consider the open cell tube shown in Fig. 5.20. The horizontal walls have a thickness t and the vertical wall has a thickness of $2t$. The boom areas are shown in the figure. We want to compute the shear flows in all the walls and the location of the shear center. For this cell,

$$I = 10Ba^2.$$

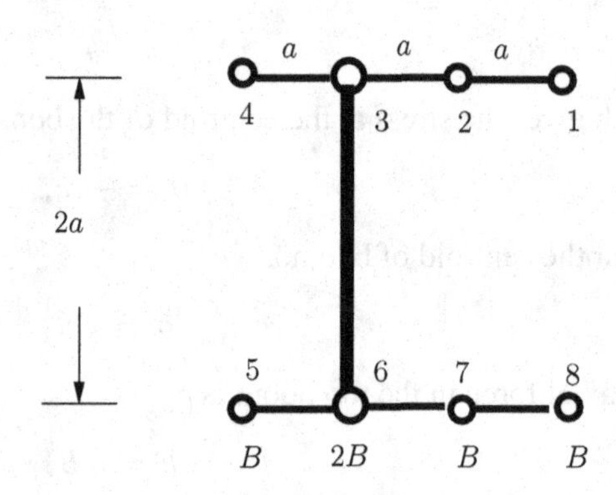

Figure 5.20 An open cell beam under the boom approximation.

Using the node numbering shown in Fig. 5.20,

$$q_{12} = -\frac{VBa}{I} = -\frac{V}{10a}, \quad q_{23} = -\frac{2V}{10a}, \quad q_{43} = -\frac{V}{10a},$$

$$q_{36} = q_{23} + q_{43} - \frac{2V}{10a} = -\frac{5V}{10a}.$$

Fig. 5.21 shows the shear flows as multiples of $V/(10a)$.

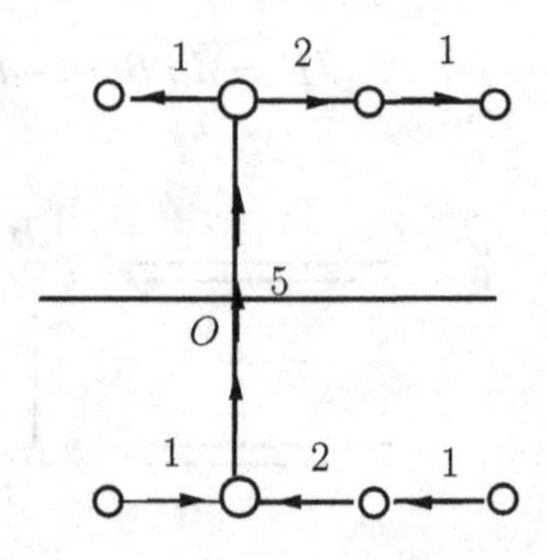

Figure 5.21 Shear flow distribution as multiples of $V/(10a)$.

To find the shear center for this beam, let us take the moment due to the shear flows about the point O on the neutral axis. This gives a clockwise moment of

$$M = 2q_{23}a^2 = 4\frac{Vta^4}{I} = \frac{4Va}{10}. \tag{5.60}$$

The shear force has to be acting through a point $2a/5$ to the left of the vertical web to create this moment.

5.4.2 Example: A Two-cell Beam under Boom Approximation

Consider the two-cell beam with a uniform wall thickness t and constant boom areas B shown in Fig. 5.22(a). The cells are numbered from left to right. Our

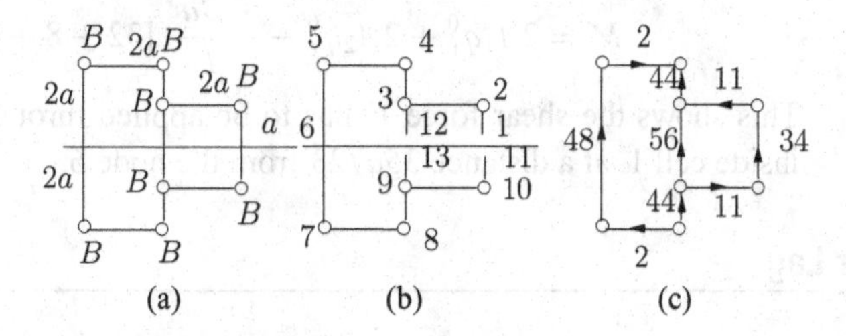

Figure 5.22 Two-cell beam: (a) geometry; (b) node numbers; and (c) shear flows in multiples of $VBa/(23I)$.

objectives are to find the shear flows in the walls and the location of the shear center. Fig. 5.22(b) shows the node numbering after applying two cuts. For this geometry we have

$$I = 20Ba^2.$$

Starting from the node 1,

$$q^v_{12} = 0, \quad q^v_{23} = -\frac{VBa}{I}, \quad q^v_{123} = 0, q^v_{34} = -\frac{2VBa}{I},$$

$$q^v_{45} = -\frac{4VBa}{I}, \quad q^v_{56} = -\frac{6VBa}{I}.$$

We can extend these to the lower half as we have done before. The areas of the cells are: $A_1 = 8a^2$ and $A_2 = 4a^2$. The constant shear flows at the cuts are found from

$$\oint q ds/t = 0,$$

for each cell. Thus,

$$12aq^0_1 - 2aq^0_2 = \frac{VBa^2}{I}[6 \times 4a + 4 \times 4a + 2 \times 2a] = \frac{44VBa^2}{I},$$

$$-2aq^0_1 + 8aq^0_2 = \frac{VBa^2}{I}[1 \times 4a] = \frac{4VBa^2}{I}.$$

Solving these equations gives

$$q^0_1 = \frac{90VBa}{23I}, \quad q^0_2 = \frac{34VBa}{23I}.$$

Adding the constant shear flows q^0 to the varying shear flows q^v, we get

$$q_{12} = \frac{34VBa}{23I}, \quad q_{23} = \frac{11VBa}{23I}, \quad q_{123} = \frac{56VBa}{23I},$$

$$q_{34} = \frac{44VBa}{23I}, \quad q_{45} = -\frac{2VBa}{23I}, \quad q_{56} = -\frac{48VBa}{23I}.$$

Here, if q_{ij} is positive, the shear flow is directed from node i to node j. Fig. 5.22(c) shows the shear flows in direction and magnitudes in multiples of $VBa/(23I)$.

In computing the location of the shear center, it is more convenient to keep q^0 and q^v separate. If we take a counterclockwise moment about the node 6, we find its total value is

$$M = 2A_1q_1^0 + 2A_2q_2^0 - \frac{VBa^3}{I}[32 + 8 + 4] = \frac{35Va}{23}.$$

This shows the shear force V has to be applied through a point that is located inside cell 1, at a distance $35a/23$ from the node 6.

5.5 Shear Lag

In the sheet-stringer construction, when adjacent stringers carry unequal axial stresses, the higher stress decreases and the lower stress increases along the length as the load is transmitted from one stringer to the other through the shearing action of the sheet. This is known as shear lag. To illustrate this, consider the sheet-stringer arrangement shown in Fig. 5.23(a). We assume all stringers have cross-

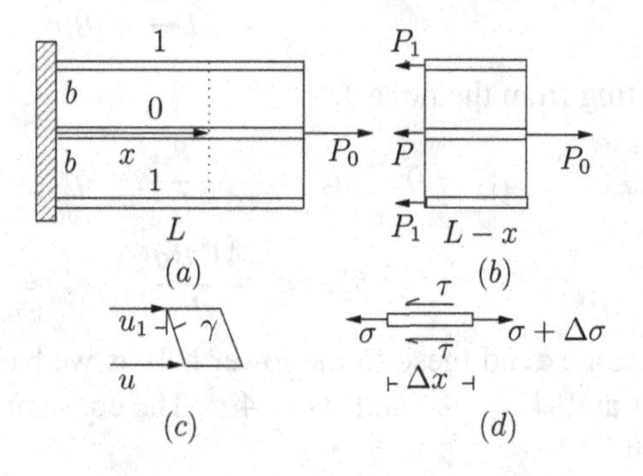

Figure 5.23 Load transmission across webs.

sectional area B, length L, and Young's modulus E, and the sheet between them has a depth b, thickness t, and shear modulus G. The middle stringer has an end load P_0. The outer stringers are load-free. The free-body diagram in Fig. 5.23(b), obtained after making a cut at x, shows the requirement for equilibrium

$$P + 2P_1 = P_0, \tag{5.61}$$

where P is the tension in the middle stringer 0 and P_1 the tension in the outer stringers 1. Denoting the axial displacements in 0 and 1 by u and u_1, respectively,

$$\epsilon = \frac{\partial u}{\partial x}, \quad \epsilon_1 = \frac{\partial u_1}{\partial x}. \tag{5.62}$$

The stress-strain relations give

$$\sigma = \frac{P}{B} = E\frac{\partial u}{\partial x}, \quad \sigma_1 = \frac{P_1}{B} = E\frac{\partial u_1}{\partial x}. \tag{5.63}$$

The sheets undergo shear strains due to the difference in the axial displacement of the stringers (see Fig. 5.23(*c*)).

$$\gamma = \frac{u - u_1}{b} = \frac{\tau}{G}, \quad \text{or} \quad \tau = \frac{G}{b}(u - u_1). \tag{5.64}$$

Equilibrium of a piece of the stringer 0 (see Fig. 5.23(*d*)) shows

$$B\Delta\sigma = 2\tau t\Delta x, \quad \text{or} \quad \frac{\partial \sigma}{\partial x} = \frac{2\tau t}{B} = \frac{2Gt}{Bb}(u - u_1). \tag{5.65}$$

Multiplying this expression by B,

$$\frac{\partial P}{\partial x} = \frac{2Gt}{b}(u - u_1). \tag{5.66}$$

Differentiating one more time and using Eq. (5.63), we have

$$\frac{\partial^2 P}{\partial x^2} = \frac{2Gt}{EBb}(P - P_1). \tag{5.67}$$

We may eliminate P_1 using Eq. (5.61), to get the differential equation for P,

$$\frac{\partial^2 P}{\partial x^2} = \frac{Gt}{EBb}(3P - P_0). \tag{5.68}$$

Defining

$$\lambda^2 = \frac{3Gt}{EBb}, \tag{5.69}$$

we rewrite the differential equation in the form

$$\frac{\partial^2 P}{\partial x^2} - \lambda^2 P = -\frac{\lambda^2}{3}P_0. \tag{5.70}$$

This equation has the solution

$$P = Ae^{-\lambda x} + Be^{\lambda x} + \frac{1}{3}P_0. \tag{5.71}$$

One boundary condition we know: $P(L) = P_0$. This leads to

$$Ae^{-\lambda L} + Be^{\lambda L} = \frac{2}{3}P_0. \tag{5.72}$$

As shown in the figure, at $x = 0$, we have $u = u_1 = 0$. From Eq. (5.66), this is equivalent to $dP/dx = 0$. This is the second boundary condition. Then,

$$A - B = 0. \tag{5.73}$$

Solving for A and B,

$$A = B = \frac{P_0}{3\cosh \lambda L}. \tag{5.74}$$

Finally,

$$P = \frac{1}{3}P_0\left[1 + 2\frac{\cosh \lambda x}{\cosh \lambda L}\right]. \tag{5.75}$$

If $\lambda x \gg 1$, $\cosh \lambda x \sim e^{\lambda x}/2$, and

$$P = \frac{1}{3} P_0 \left[1 + 2e^{-\lambda(L-x)} \right]. \tag{5.76}$$

This expression shows the exponential decay of the force P as we move from the loaded end $x = L$. The forces in the outer stringers exponentially increase as x decreases.

If the left end is located at a "cut-out" where the stringer abruptly stops, the second boundary condition will be $P = 0$ at $x = 0$. Then

$$A + B = -\frac{1}{3} P_0. \tag{5.77}$$

Solving for A and B, we get

$$P = \frac{1}{3} P_0 \left[1 + \frac{2 \sinh \lambda x - \sinh \lambda (L - x)}{\sinh \lambda L} \right]. \tag{5.78}$$

The increased load in the outer stringers has to be taken into account while introducing cut-outs in fuselages and wings.

5.6 Combined Bending and Torsion

When a shear force V does not pass through the shear center, there will be bending and twisting of the beam. To analyze this case, we may consider two approaches: (a) find the shear center and separate the problem into a bending problem with the shear force through the shear center, and into a torsion problem with a torque equal to the shear force times the distance between the shear center and the actual point of application of the load, or (b) find the shear flow distribution that introduces a rate of twist and zero moment about the point of application of the shear force. In the following example, we consider case (b).

5.6.1 Example: Single-cell Beam

The single-cell beam shown in Fig. 5.24 has boom areas B and web widths $2a$. The wall thickness of the left wall is $3t$ and all other walls have thicknesses of t. Assume a uniform shear modulus of G. A shear force V is applied vertically through the left wall. The second moment of the area for the booms is

$$I = 4Ba^2.$$

After introducing a cut in the right wall,

$$q_{12}^v = 0, \quad q_{23}^v = -\frac{VBa}{I}, \quad q_{34}^v = -\frac{2VBa}{I}.$$

We extend these antisymmetrically to the lower half of the beam. A shear flow of q^0 is added to close the cut. The moment about the point 4 is found as

$$M = 8a^2 q^0 - 2 \times 2a^2 \frac{VBa}{I}.$$

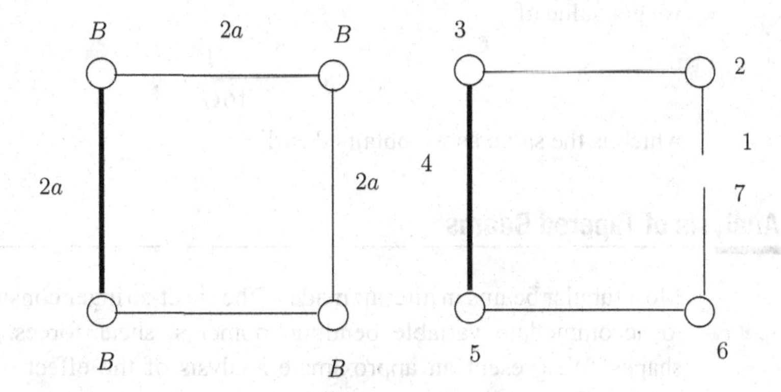

Figure 5.24 Single-cell beam: geometry, node numbers, and shear flows in multiples of $VBa/(23I)$.

This moment is zero as the shear force passes through the point 4. Then

$$q^0 = \frac{VBa}{2I}.$$

The rate of twist is obtained from

$$\alpha = \frac{1}{2AG}\oint \frac{q\,ds}{t},$$

as

$$\alpha = -\frac{V}{16Ga^2t}.$$

As we have q positive in the counterclockwise direction, the negative sign in front of α implies a clockwise rate of twist.

If the shear force were to pass through the shear center, there will not be any rate of twist. In this case we compute q^0 using

$$\oint \frac{q\,ds}{t} = 0, \quad q^0\left[\frac{6a}{t} + \frac{2a}{3t}\right] = \frac{VBa}{I}\left[\frac{4a}{t} + \frac{4a}{3t}\right].$$

This gives

$$q^0 = \frac{4}{5}\frac{VBa}{I}.$$

The moment about the node 4 is now obtained as

$$M = 2Aq^0 - \frac{VBa}{I}4a^2, \quad M = \frac{3Va}{5}.$$

Thus, the shear center for this beam is $3a/5$ to the right of node 4. In the original problem, the shear force passes through node 4. This is equivalent to a force of V through the shear center and a counterclockwise torque

$$T = -3Va/5.$$

Using the formula

$$\alpha = \frac{T}{4A^2G}\oint \frac{ds}{t},$$

we get value of

$$\alpha = -\frac{V}{16Ga^2t},$$

which is the same as we obtained earlier.

5.7 Analysis of Tapered Beams

Most tubular beams in aircraft made of the sheet-stringer construction are tapered to accommodate variable bending moments, shear forces, and aerodynamic shapes. We present an approximate analysis of the effect of taper in the following. To simplify the analysis, we assume the beam is of uniform width in the z-direction, is symmetric about the z-axis, and tapers in the y-direction. We further assume the tapered booms will converge to a single point if extended.

5.7.1 Quadrilateral Shear Panels

In the case of tapered sheet-stringer construction, we encounter quadrilateral sheets (shear panels) pictured in Fig. 5.25. We make the assumption that the tapered stringers meet at a common point O. With the node numbers shown in

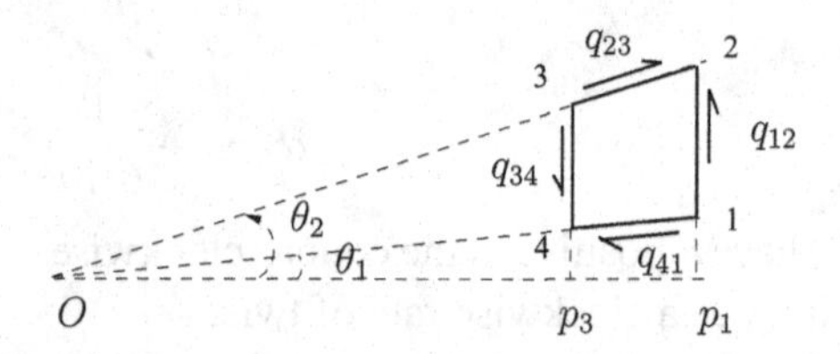

Figure 5.25 Quadrilateral shear panel with assigned nodes.

the figure, we may denote the vertical side lengths by h_{ij}, where i and j are the node numbers. Taking the moment about O,

$$q_{12}h_{12}p_1 = q_{34}h_{34}p_3, \quad q_{34} = q_{12}\frac{h_{12}p_1}{h_{34}p_3} = q_{12}\frac{p_1^2}{p_3^2}, \tag{5.79}$$

where we have used a property of similar triangles. Given q_{12}, this formula can be used to find q_{34} along any vertical cut and it is inversely proportional to the distance to O.

By balancing the horizontal forces,

$$q_{23}h_{23}\cos\theta_2 = q_{41}h_{41}\cos\theta_1, \quad q_{23} = q_{41}. \tag{5.80}$$

As we will see, q_{23} and q_{41} are average shear flows – their local values vary from node to node. Balancing the vertical forces,

$$q_{41}h_{41}\sin\theta_1 + q_{34}h_{34} = q_{12}h_{12} + q_{23}h_{23}\sin\theta_2, \tag{5.81}$$

where the lengths h_{ij} are related in the form

$$h_{34} - h_{12} = h_{41}\sin\theta_1 - h_{23}\sin\theta_2. \tag{5.82}$$

Using this, the vertical force balance gives

$$q_{41} = q_{12}\frac{p_1}{p_3}. \tag{5.83}$$

This expression shows that the average value of the shear flow depends on the distances of the end nodes from O.

5.7.2 Shear Flow due to Bending in Tapered Beams

Let us consider a beam consisting of N booms, all converging to a line through point O in the x, y-plane. The strain formula

$$\epsilon = -\kappa y \tag{5.84}$$

gives the change of length in the x-direction for a boom segment of length unity. If the boom segment is oriented along a unit vector

$$\boldsymbol{n} = n_x \boldsymbol{i} + n_y \boldsymbol{j}, \tag{5.85}$$

then the boom elongation for the ith boom is $-\kappa y_i / n_x$, where y_i is the coordinate of the boom. The x-axis forms the neutral axis. The stress in the boom is $-\kappa E y_i / n_x$. This creates a force

$$P_i = -\kappa E B_i y_i / n_x, \tag{5.86}$$

where B_i is the boom area. Its x-component will be

$$P_{ix} = -\kappa E B_i y_i. \tag{5.87}$$

Using

$$M = -\sum P_{ix} y_i, \tag{5.88}$$

we find

$$\kappa = \frac{M}{EI}, \quad I = \sum B_i y_i^2, \tag{5.89}$$

and

$$P_{ix} = \frac{M B_i y_i}{I}, \quad P_i = \frac{M B_i y_i}{I n_x}. \tag{5.90}$$

From Fig. 5.26, the equilibrium equation relating bending moment and shear force is obtained as

$$\frac{dM}{dx} = -V - \sum_{j=1}^{N} P_{jy}, \tag{5.91}$$

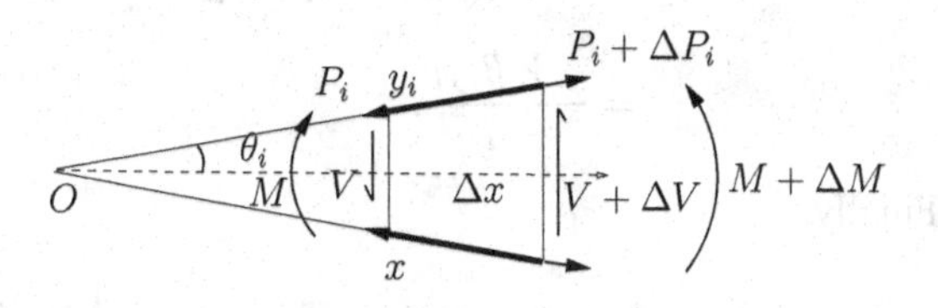

Figure 5.26 Element of a tapered beam.

where $P_{jy} = P_{ix} \tan\theta_i$ is the y-component of the force in the jth boom. As shown in Fig. 5.27, a force balance on the boom segment of length $\Delta x/n_x$ shows

$$\Delta P_i = -(q_1 + q_2 + q_3)\Delta x/n_x, \quad q_1 + q_2 + q_3 = -\frac{dP_{ix}}{dx}. \tag{5.92}$$

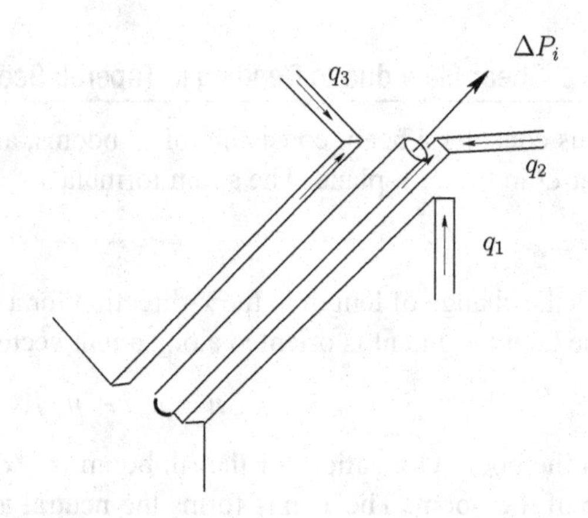

Figure 5.27 Element of a boom at the intersection of three webs.

Using Eq. (5.90) and assuming B_i are constants, we find

$$-\frac{dP_{ix}}{dx} = \frac{dM}{dx}\frac{B_i y_i}{I} + \frac{MB_i}{I}\frac{dy_i}{dx} - \frac{MB_i y_i}{I^2}\frac{dI}{dx}. \tag{5.93}$$

Our assumption that all the booms converge to a point O implies

$$y_i = x\tan\theta_i. \tag{5.94}$$

Then,

$$\begin{aligned}
-\frac{dP_{ix}}{dx} &= \frac{dM}{dx}\frac{B_i y_i}{I} + \frac{MB_i}{I}\tan\theta_i - \frac{MB_i y_i \sum 2B_j y_j \tan\theta_j}{I^2}, \\
&= \left[-V - \sum P_{jx}\tan\theta_j\right]\frac{B_i y_i}{I} + \frac{MB_i y_i}{Ix} - 2\frac{MB_i y_i}{Ix}, \\
&= -\frac{VB_i y_i}{I} + \frac{MB_i y_i}{I^2}\sum B_j y_j \tan\theta_j - \frac{MB_i y_i}{Ix}, \\
&= -\frac{VB_i y_i}{I} + \frac{MB_i y_i}{I^2}\frac{I}{x} - \frac{MB_i y_i}{Ix}, \\
&= -\frac{VB_i y_i}{I}.
\end{aligned} \tag{5.95}$$

Finally

$$q_1 + q_2 + q_3 = -\frac{VB_i y_i}{I}. \tag{5.96}$$

FURTHER READING

Curtis, H. D., *Fundamentals of Aircraft Structural Analysis*, Irwin (1997).
Megson, T. H. G., *Aircraft Structures for Engineering Students*, Butterworth-Heinemann (2007).
Rivello, R. M., *Theory and Analysis of Flight Structures*, McGraw-Hill (1969).
Sun, C. T., *Mechanics of Aircraft Structures*, John Wiley (2006).

EXERCISES

5.1 A simply supported beam of length $4a$ is subjected a concentrated moment of M_0 and a distributed load of $M_0/(2a^2)$, as shown in Fig. 5.28. Using singularity functions, obtain the shear force and the bending moment diagrams for this loading. Indicate the locations of their maximum values.

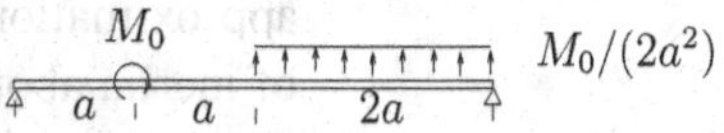

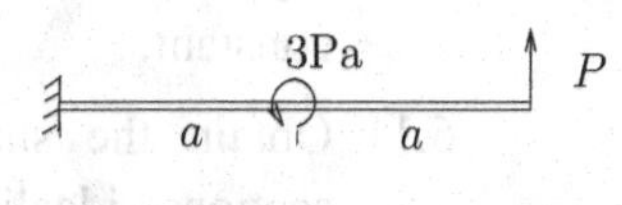

Figure 5.28 Simply supported beam.

5.2 A cantilever beam is loaded with a tip load P and a concentrated moment 3 Pa. Draw the shear force and bending moment diagrams using the singularity functions and indicate the location of their maximum values.

5.3 The beam in Fig. 5.28 has bending stiffness EI. Calculate its slope and deflection.

5.4 The beam in Fig. 5.29 has bending stiffness EI. Compute its tip deflection using the complementary energy method.

3Pa P a a

Figure 5.29 Cantilever beam.

5.5 A cantilever beam is fixed to the wall at the left end, $x = 0$. In $0 < x < a$, it has a bending stiffness of $2EI$ and in $a < x < 2a$, it has EI. Express the inverse of the stiffness distribution using the step function and obtain the beam's tip deflection and slope due to a concentrated force P at its tip.

5.6 Under the thin-wall assumption find the centroid of the uniformly thin cross-section with thickness t shown in Fig. 5.30. Also compute the second moments of the area, I_{yy}, I_{zz}, and I_{yz}.

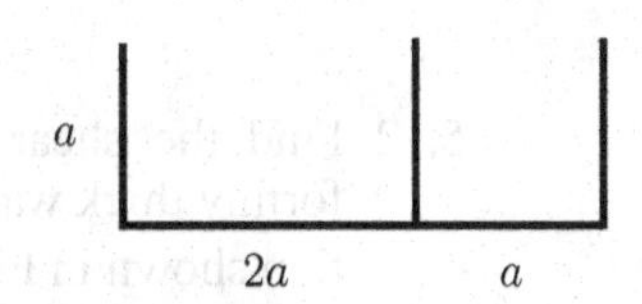

Figure 5.30 An open cross-section.

5.7 The channel section shown in Fig. 5.31 is subjected to a bending moment of 300 Nm. If $a =$ 10 cm and the thickness $t =$ 5 mm, obtain the maximum tensile and the maximum compressive stresses and their locations.

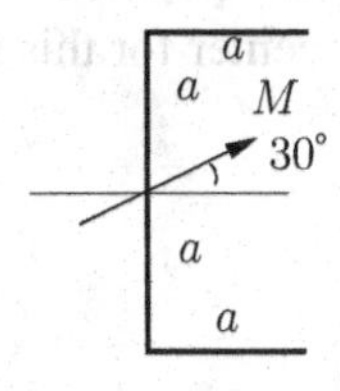

Figure 5.31 A channel section.

5.8 The uniformly thick slit rectangle shown in Fig. 5.32 is subjected to a vertical shear force V. Sketch the shear flow distribution around the walls. Obtain the location of the shear center.

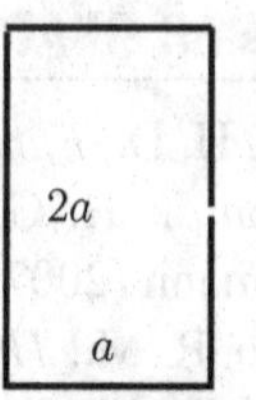

Figure 5.32 A slit rectangle.

5.9 For the T-section shown in Fig. 5.33, the flange has a thickness t and the web has $2t$. Compute the centroid of this section under the thin-wall approximation. Find the moment of inertia about the horizontal axis through the centroid. If a vertical shear force of V acts at this section, directed upward, what are the maximum shear stresses in the flange and the web?

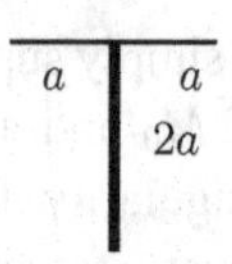

Figure 5.33 A T-section.

5.10 Obtain the shear center of the unsymmetrical section shown in Fig. 5.34. The wall thickness is a constant.

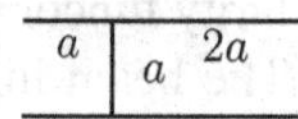

Figure 5.34 An unsymmetric I-section.

5.11 Obtain the shear flows in open section, idealized shear panels (Fig. 5.35) if they have a constant thickness t and the boom areas are: $B_1 = B_3 = B_4 = B_6 = B$ and $B_2 = B_5 = 2B$. The applied shear force is V.

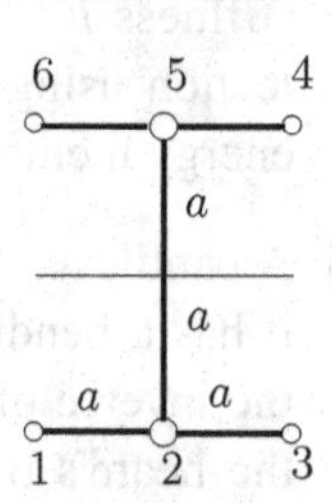

Figure 5.35 An I-beam with booms.

5.12 Find the shear flows in the uniformly thick walls of the open section shown in Fig. 5.36. The boom areas are B and the applied shear force is V. Using the shear flows, compute the location of the shear center for this section.

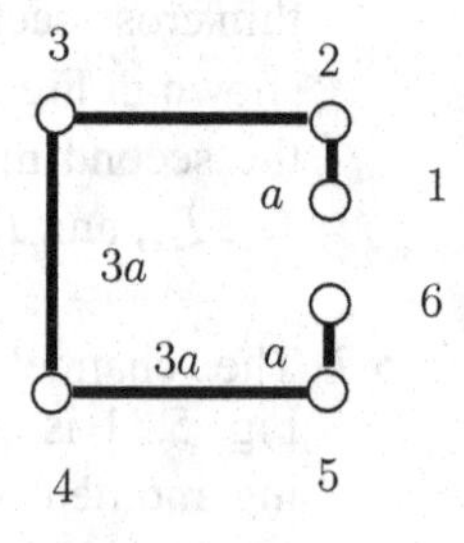

Figure 5.36 An open section.

5.13 A uniform thickness, closed cell is shown in Fig. 5.37. Under an applied shear force of V, compute the shear flows in all the walls and find the shear center for this section. All booms have the same area, B.

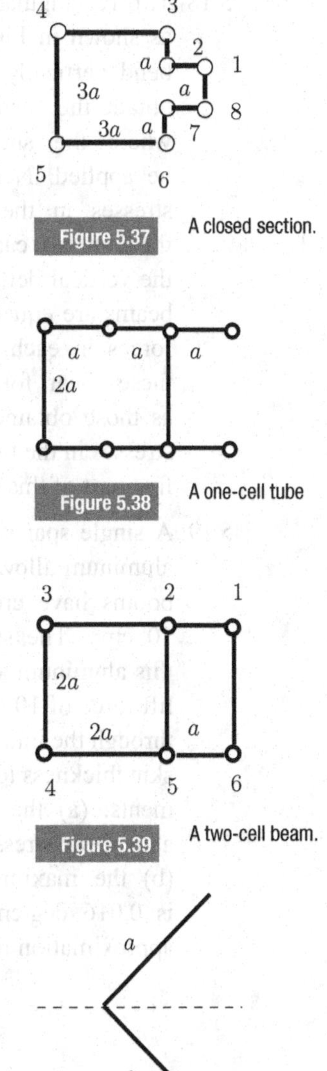

Figure 5.37 A closed section.

5.14 The cross-section shown in Fig. 5.38 has a vertical shear of force V. The wall thickness is t. (a) Compute the moment of inertia about the axis of symmetry; (b) make a cut and compute the shear flow distribution; (c) find the shear flow at the cut; and (d) find the shear center for this section.

Figure 5.38 A one-cell tube

5.15 For the uniformly thick two-cell beam shown in Fig. 5.39, the boom areas are B. Compute the shear flows in all the walls and find the shear center.

Figure 5.39 A two-cell beam.

5.16 An angle section beam is subjeced to a vertical shear force V. The two walls are at an angle of 45° with the x-axis. Assuming uniform thickness t, obtain the second moment of area I_{xx} and the shear flow distribution along the walls. Where is the maximum shear flow? Find the shear center for this section.

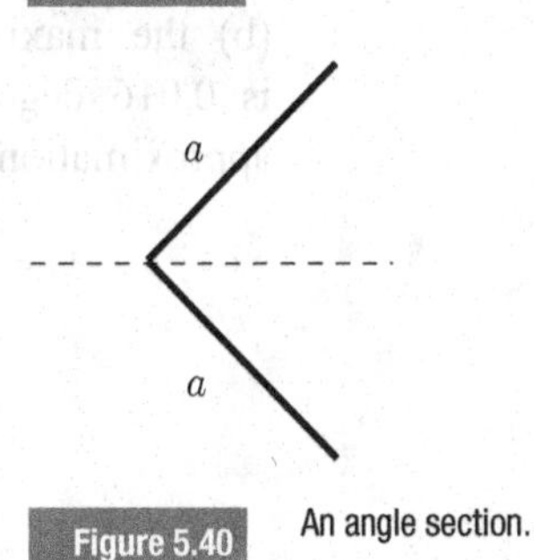

Figure 5.40 An angle section.

5.17 A beam cross-section with three closed cells is shown in Fig. 5.41. All walls are 5 mm thick. The cell areas are: $A_1 = 140$ cm^2, $A_2 = 240$ cm^2, and $A_3 = 96$ cm^2. The wall lengths are: $l_{24} = 20$ cm, $l_{15} = 12$ cm, and $l_{23} = 18$ cm. All booms have cross-sectional area of 2 cm^2. The applied shear force through the shear center is 200 kN. Compute the shear stresses in all the walls. Find the location of the shear center. Use the boom approximation to neglect the bending stresses in the panels.

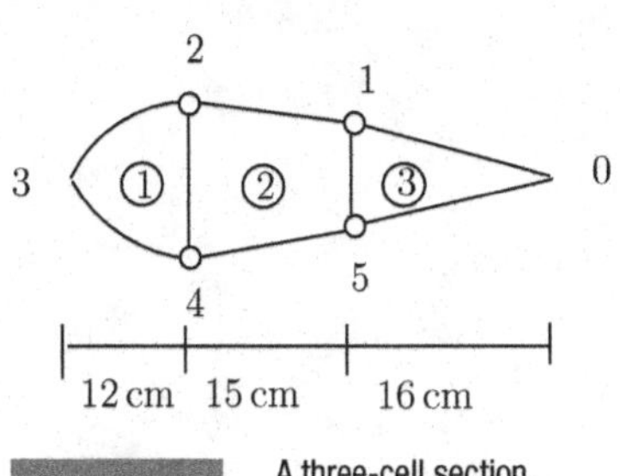

Figure 5.41 A three-cell section.

5.18 Two rectangular section beams of width t are connected by a thin strip as shown in Fig. 5.42. They carry a net vertical shear force V. If they bend vertically without twisting, obtain the location on the strip where the shear force has to be applied. Neglect the bending stresses in the strip. Assuming these are two cantilever beams and the vertical deflections of the two beams are equal, obtain the shear forces in each beam. Show that these shear forces are the same as those obtained from the shear stresses in the two beams from the first part of this problem.

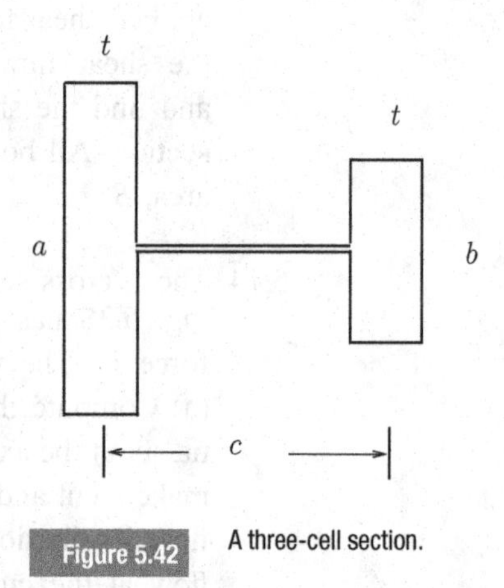

Figure 5.42 A three-cell section.

5.19 A single spar supersonic tail section, shown in Fig. 5.43, is made of an aluminum alloy. The skin and the spar web have the same thickness. All booms have cross-sectional area 10 cm^2. The shear modulus for this aluminum alloy is 27 GPa. A lift force of 10 MN acts vertically through the mid-chord. Design the skin thickness to meet the requirements: (a) the maximum allowable shear stress is 200 MPa and (b) the maximum rate of twist is 0.016 deg/cm. Use the boom approximation for the calculation.

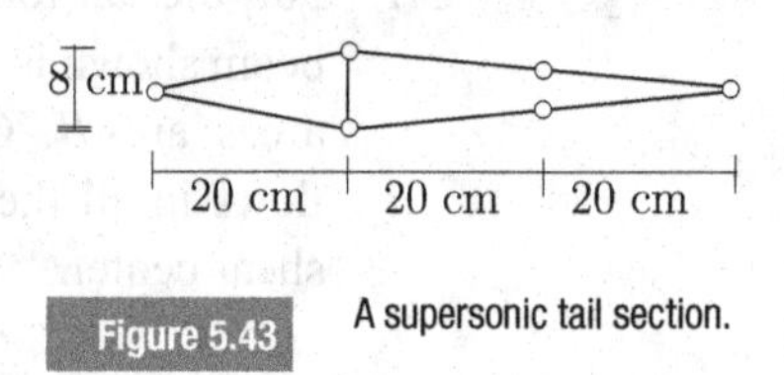

Figure 5.43 A supersonic tail section.

6 Stability of Structures

So far in our studies, we have been concerned with the equilibrium of structures. Once equilibrium is established, the next level involves ascertaining the stability of the equilibrium state. In this chapter we consider only static stability; dynamic stability analysis, which takes into account the inertia forces and oscillations, is also an important topic. A simple way to demonstrate the concept of stability is to consider a ruler in the vertical position, as shown in Fig. 6.1. Using statics, we

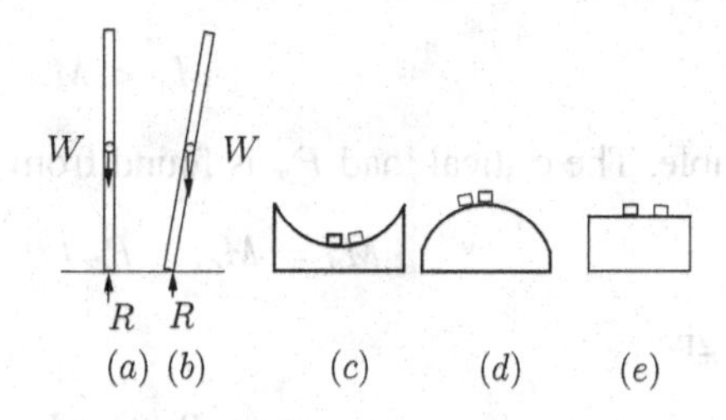

Figure 6.1 Equilibrium states and perturbed states.

can balance its weight by applying a reaction $R = W$ as seen in (a). However, a slight perturbation from the ideal vertical position creates a moment, as shown in (b), and the ruler falls to the side.

In Fig. 6.1, (c), (d), and (e) show stable, unstable, and neutral equilibrium of frictionless blocks on surfaces. In (c), when the block is moved from the equilibrium state at the bottom of the concave surface, it returns. In (d), the block does not return. In (e) we have neutral stability – the block stays in the perturbed state.

In order to investigate the stability, first we establish possible equilibrium states and then perturb the system slightly to see if it would come back to the equilibrium state nearest to it.

6.0.3 Spring-supported Vertical Bar

Consider a vertical rigid bar with a spring support as shown in Fig. 6.2. From the vertical equilibrium position if we disturb the bar by a small angle θ, the disturbing moment is

$$M_d = PL\theta, \tag{6.1}$$

and the restoring moment due to the stretch in the spring is

$$M_r = kL^2\theta. \tag{6.2}$$

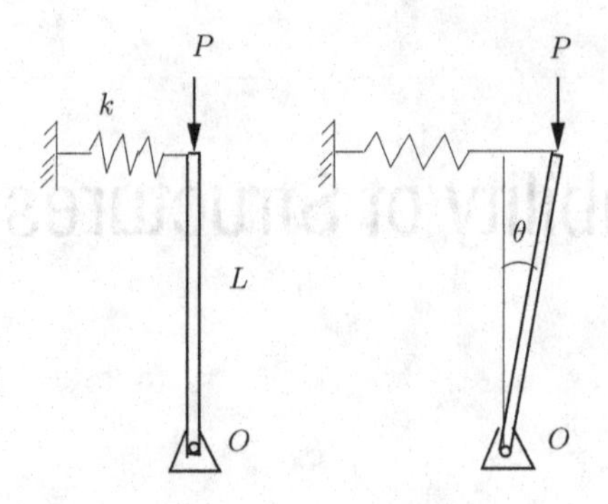

Figure 6.2 Spring-supported vertical bar.

Note, for small angles, $\sin\theta \approx \theta$ and $\cos\theta \approx 1$.

If

$$M_d > M_r, \tag{6.3}$$

the bar is unstable, and if

$$M_d < M_r, \tag{6.4}$$

it is stable. The critical load P_{cr} is found from

$$M_d = M_r, \quad P_{cr} L\theta = kL^2\theta, \tag{6.5}$$

which gives

$$P_{cr} = kL. \tag{6.6}$$

In this example the critical load is proportional to the spring stiffness.

6.1 Buckling of a Beam

Consider a beam with bending stiffness EI and of length L, simply supported at both the ends. If we subject it to an axial compressive load of P, it will be in equilibrium if P is less than the yield load $\sigma_y A$, where σ_y is the yield stress in compression and A is the cross-sectional area of the beam. We have to allow one of the simple supports to move horizontally as the load is increased. Let us assume the right end is a sliding support. We can disturb the equilibrium by deflecting the beam in the vertical direction. This is shown in Fig. 6.3. The bending moment at x due to the eccentricity of the load is

$$M = -Pv. \tag{6.7}$$

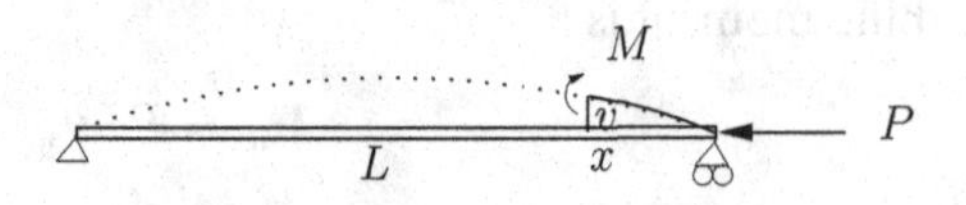

Figure 6.3 Buckling of a beam.

Using this in the deflection equation

$$EIv'' = M, \tag{6.8}$$

we find

$$v'' + \lambda^2 v = 0, \quad \lambda^2 = \frac{P}{EI}. \tag{6.9}$$

The boundary conditions are

$$v(0) = 0, \quad v(L) = 0. \tag{6.10}$$

The differential equation and the boundary conditions are satisfied by the trivial solution $v(x) = 0$, which is the undisturbed state of the beam. Using

$$v(x) = C \cos \lambda x + D \sin \lambda x, \tag{6.11}$$

which satisfies the differential equation, in the boundary conditions, we get

$$C = 0, \quad D \sin \lambda L = 0. \tag{6.12}$$

As C and D, both, cannot be zero,

$$\sin \lambda L = 0, \quad \lambda = \frac{n\pi}{L}, \quad n = 1, 2, \ldots. \tag{6.13}$$

The critical loads are given by

$$P_{cr} = \frac{n^2 \pi^2 EI}{L^2}. \tag{6.14}$$

The lowest critical load is

$$P_{cr} = \frac{\pi^2 EI}{L^2}, \tag{6.15}$$

which corresponds to $n = 1$. These types of problems, which allow a trivial solution $v(x) = 0$ and a sequence of solutions for discrete values of a parameter λ, are called differential eigenvalue problems, λ being the eigenvalue. For $n = 1$, we have

$$v(x) = D \sin \frac{\pi x}{L}, \tag{6.16}$$

and for any value of n

$$v(x) = D \sin \frac{n\pi x}{L}. \tag{6.17}$$

These shapes are called the buckling modes and the amplitude D remains an unknown. As our solution is based on small deflections of beams, D has to be kept small.

The first mode,

$$v(x) = D \sin \pi x / L, \tag{6.18}$$

has zero amplitude only at the end points. The second mode,

$$v(x) = D \sin 2\pi x / L, \tag{6.19}$$

is zero at the point $x = L/2$. This point is called a node. So, we have one node for the second mode. For the nth mode there will be $(n - 1)$ nodes. Thus, we can identify the mode based on the number of nodes.

This derivation was done by Euler and it is known as Euler's theory of buckling of beams or columns.

6.1.1 Critical Stresses

At critical buckling load P_{cr} the compressive stress in the beam is

$$\sigma_{cr} = -P_{cr}/A, \tag{6.20}$$

where A is the area of the beam cross-section. Introducing the radius of gyration ρ as in

$$I = A\rho^2, \tag{6.21}$$

we can write

$$\sigma_{cr} = -\pi^2 E \left(\frac{\rho}{L}\right)^2. \tag{6.22}$$

The ratio L/ρ is called the slenderness ratio of the beam. The critical stress is inversely proportional to the square of the slenderness ratio. For short beams (low slenderness ratio) this formula, which is known as the Euler hyperbola, gives high values for the critical stress – exceeding the yield stress of the material. We may plot critical stress against the slenderness ratio as shown in Fig. 6.4.

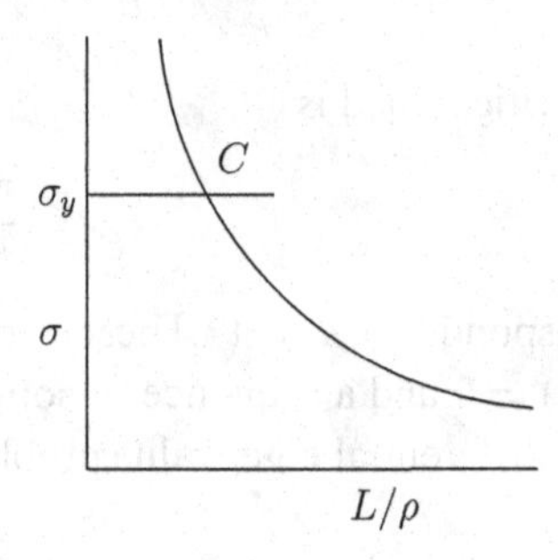

Figure 6.4 Domain of safe stresses in the stress versus slenderness ratio plot.

Below the Euler hyperbola and the horizontal line indicating the yield stress, any state of stress is safe, except near the corner C, where plastic deformation and buckling occur simultaneously. As our derivation of the Euler buckling load does not take into account plastic strains in bending, the corner area should not be considered as safe. There are improved theories that smooth the transition between buckling and plastic flow.

6.2 Bending under an Eccentric Load

Consider a beam loaded in compression, Fig. 6.5, with an eccentrically applied load. Unlike the Euler buckling, the state of the beam as a straight line does not satisfy equilibrium.

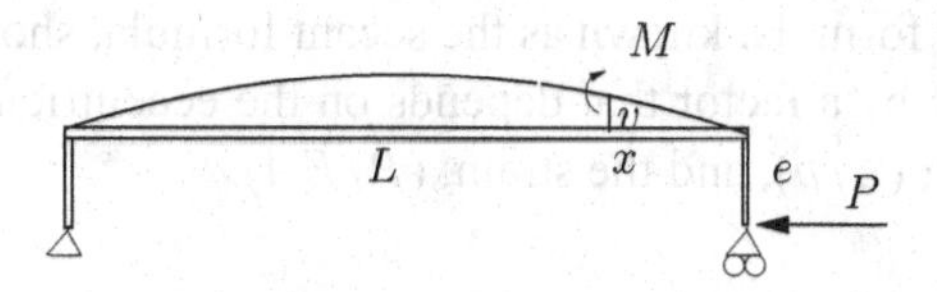

Figure 6.5 Eccentrically loaded beam.

The bending moment at x is

$$M = -P(v + e), \tag{6.23}$$

and the moment-curvature relation gives

$$EIv'' + Pv = -Pe. \tag{6.24}$$

The boundary conditions are

$$v(0) = 0, \quad v(L) = 0. \tag{6.25}$$

Using a particular solution and a complementary solution of the differential equation, the general solution is

$$v(x) = C\cos\lambda x + D\sin\lambda x - e, \tag{6.26}$$

where $\lambda^2 = P/(EI)$. The boundary conditions are satisfied by

$$C = e, \quad D = e\frac{1-\cos\lambda L}{\sin\lambda L} = e\frac{\sin\lambda L/2}{\cos\lambda L/2}. \tag{6.27}$$

Then

$$\begin{aligned} v(x) &= e\left[\cos\lambda x + \frac{\sin\lambda x\sin\lambda L/2}{\cos\lambda L/2} - 1\right], \\ &= e\left[\frac{\cos\lambda(x - L/2)}{\cos\lambda L/2} - 1\right]. \end{aligned} \tag{6.28}$$

The maximum deflection is at $x = L/2$, with the value

$$v_{max} = e\left[\frac{1}{\cos\lambda L/2} - 1\right]. \tag{6.29}$$

As the applied load P is increased from 0, the value of λ increases and so does the maximum deflection. Finally, when $\lambda L/2 = \pi/2$, v_{max} goes to infinity. The value of P, when v_{max} goes to infinity, is exactly the Euler buckling load.

Using Eq. (6.29), we find

$$M_{max} = Pe\sec\left(\frac{L}{2}\sqrt{\frac{P}{EI}}\right). \tag{6.30}$$

The maximum compressive stress

$$\begin{aligned} \sigma_{max} &= \frac{P}{A} + \frac{MC}{I}, \\ &= \frac{P}{A}\left[1 + \frac{ec}{\rho^2}\sec\left(\frac{L}{2\rho}\sqrt{\frac{P}{EA}}\right)\right]. \end{aligned} \tag{6.31}$$

This formula, known as the secant formula, shows the amplification of the stress P/A by a factor that depends on the eccentricity ratio, (ec/ρ^2), the slenderness ratio, (L/ρ), and the strain, (P/EA).

6.3 Buckling of Imperfect Beams

So far we have assumed the beams in question are perfectly straight. In reality, a beam may have preexisting imperfections in the form of small deviations of the neutral axis from the ideal straight line. In other words, we may assume there is a deflection $v_0(x)$ of the neutral axis before the axial load is imposed. This is shown in Fig. 6.6. If the elastic deflection $v(x)$ is superposed on the existing, nonelastic deflection v_0, the change in curvature is v''.

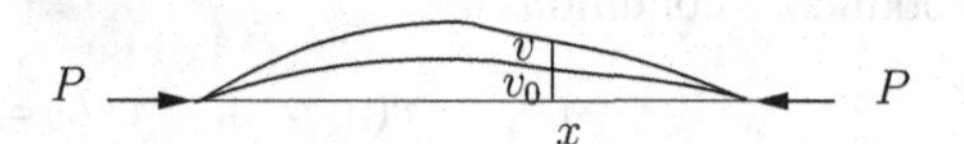

Figure 6.6 Buckling of an imperfect beam.

The equation of equilibrium is obtained as

$$EIv'' + Pv = -Pv_0. \tag{6.32}$$

The unknown initial deviation v_0 can be expanded in a Fourier series as

$$v_0(x) = A_1 \sin\frac{\pi x}{L} + A_2 \sin\frac{2\pi x}{L} + A_3 \sin\frac{3\pi x}{L} + \cdots, \tag{6.33}$$

where the coefficients are kept as unknowns. Assuming a solution for v of the form

$$v(x) = B_1 \sin\frac{\pi x}{L} + B_2 \sin\frac{2\pi x}{L} + B_3 \sin\frac{3\pi x}{L} + \cdots, \tag{6.34}$$

and substituting it in the equilibrium equation, we can equate coefficients of the sin-terms. This way, we find

$$B_1\left(\frac{\pi^2 EI}{PL^2} - 1\right) = A_1, \quad B_2\left(\frac{4\pi^2 EI}{PL^2} - 1\right) = A_2,$$
$$B_3\left(\frac{9\pi^2 EI}{PL^2} - 1\right) = A_3, \ldots \tag{6.35}$$

From these,

$$B_1 = \frac{A_1}{\frac{P_{cr}}{P} - 1}, \quad B_2 = \frac{A_2}{\frac{4P_{cr}}{P} - 1}, \quad B_3 = \frac{A_3}{\frac{9P_{cr}}{P} - 1}, \ldots, \tag{6.36}$$

where

$$P_{cr} = \frac{\pi^2 EI}{L^2}, \tag{6.37}$$

is the Euler buckling load. As in the case of eccentric loading, the amplitudes, B_i, of the elastic deflection begin to increase as the load is increased. Finally, when $P = P_{cr}$, it becomes unbounded.

Neglecting the higher order Fourier coefficients, the elastic deflection at the midpoint of the beam

$$\delta = v(L/2) = B_1 = A_1/(P_{cr}/P - 1). \tag{6.38}$$

This can be rearranged to get

$$\delta = P_{cr}\frac{\delta}{P} - A_1. \tag{6.39}$$

A plot of δ versus δ/P will have P_{cr} as its slope. This plot is called the Southwell plot and it provides a nondestructive means of obtaining the critical load.

6.4 Cantilever Beam

A cantilever beam subjected to a compressive load shown in Fig 6.7, can be analyzed as follows.

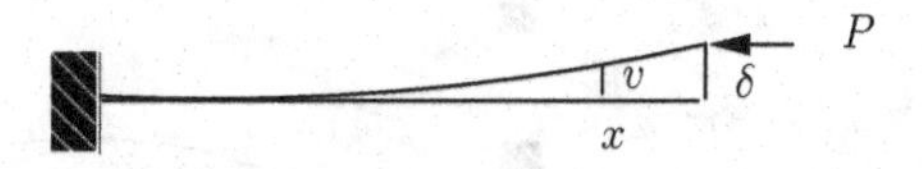

Figure 6.7 Buckling of a cantilever beam.

If the tip deflection

$$\delta = v(L), \tag{6.40}$$

the bending moment at a point x is given by

$$M = P(\delta - v), \tag{6.41}$$

and the equilibrium equation is

$$EIv'' = P(\delta - v), \tag{6.42}$$

with the boundary conditions, $v(0) = 0$, $v'(0) = 0$. The solution of the equilibrium equation can be found as

$$v(x) = \delta + A\cos\lambda x + B\sin\lambda x, \quad \lambda^2 = \frac{P}{EI}. \tag{6.43}$$

The conditions $v(0) = 0$ and $v'(0) = 0$ give

$$A = -\delta, \quad B = 0. \tag{6.44}$$

So far we have

$$v(x) = \delta(1 - \cos\lambda x). \tag{6.45}$$

To satisfy $v(L) = \delta$, we require

$$\cos\lambda L = 0. \tag{6.46}$$

Then

$$\lambda L = \pi/2, 3\pi/2, 5\pi/2, \ldots. \tag{6.47}$$

The critical load is obtained as

$$P_{cr} = \frac{\pi^2 EI}{4L^2}, \tag{6.48}$$

corresponding to the lowest value of λ.

The mode shapes are given by

$$v(x) = \delta(1 - \cos \lambda x/L). \tag{6.49}$$

6.5 Propped Cantilever Beam

A cantilever beam with a support as shown in Fig. 6.8 is called a propped cantilever. This can also be referred to as a pinned-clamped beam (a). The forces acting on this beam are shown in the free-body diagram (b).

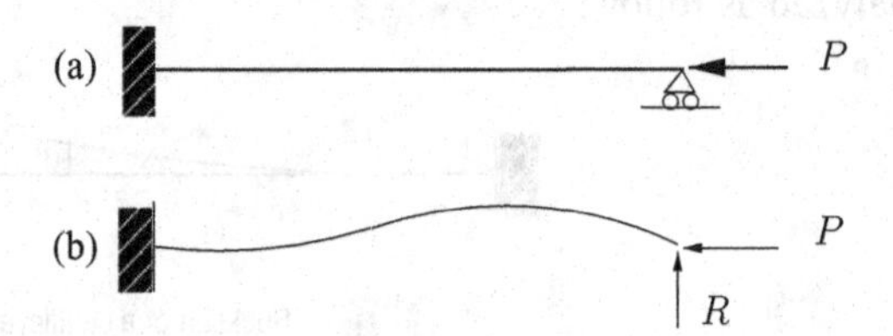

Figure 6.8 Buckling of a propped cantilever beam.

With the unknown reaction R at $x = L$, the bending moment at a section x is given by

$$M = R(L - x) - Pv. \tag{6.50}$$

The equilibrium equation

$$EIv'' + Pv = R(L - x), \tag{6.51}$$

has the solution

$$v(x) = \frac{R}{P}(L - x) + A \cos \lambda x + B \sin \lambda x, \quad \lambda^2 = \frac{P}{EI}. \tag{6.52}$$

The boundary conditions are

$$v(0) = 0, \quad v'(0) = 0, \quad v(L) = 0. \tag{6.53}$$

From the conditions at $x = 0$, we get

$$A = -\frac{RL}{P}, \quad B = \frac{R}{P\lambda}. \tag{6.54}$$

With these constants

$$v(x) = \frac{RL}{P}\left[1 - \frac{x}{L} - \cos \lambda x + \frac{\sin \lambda x}{\lambda L}\right]. \tag{6.55}$$

The last condition, $v(L) = 0$, gives the characteristic equation for λ,

$$\tan \lambda L = \lambda L. \tag{6.56}$$

The smallest root of this equation as shown in Fig. 6.9, is

$$\lambda L = 4.4934, \quad P_{cr} = \frac{20.19EI}{L^2} = \frac{2.046\pi^2 EI}{L^2}. \tag{6.57}$$

The mode shapes are given by Eq. (6.55), for different values of λL.

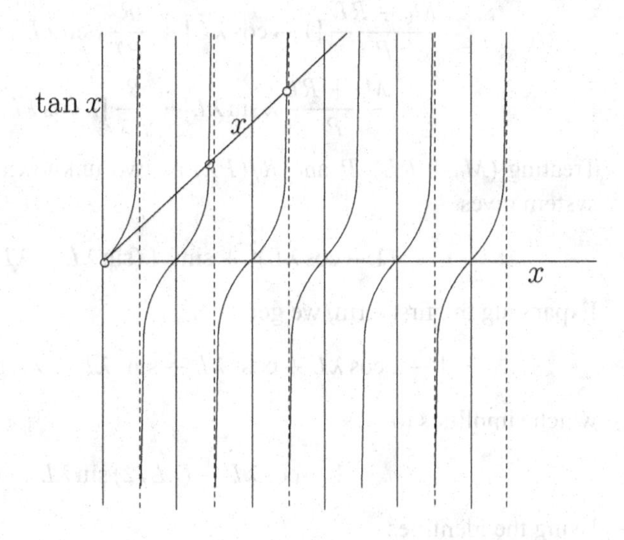

Figure 6.9 A sketch of the location of the roots of the equation $\tan x = x$, vertical lines are spaced at $\pi/2$.

6.6 Clamped-clamped Beam

Let us consider a beam with both of its ends clamped. As shown in Fig. 6.10, the free-body diagram shows a vertical force R and a moment M_0 when the clamp is released at the end $x = L$. The differential equation of equilibrium is

$$EIv'' + Pv = M_0 + R(L - x). \tag{6.58}$$

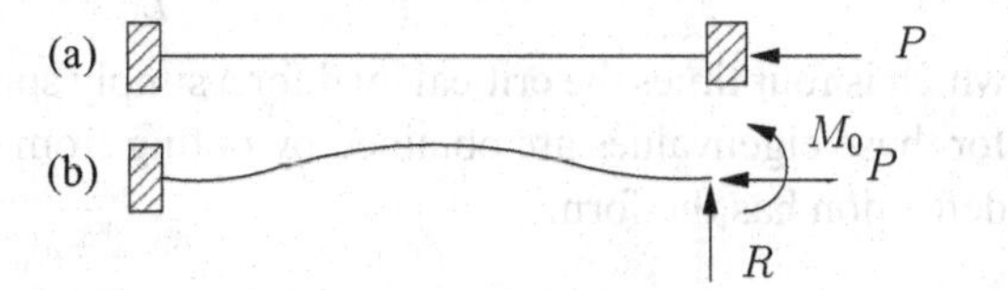

Figure 6.10 A clamped-clamped beam (a) and its free-body diagram (b).

The solution of this equation is

$$v(x) = \frac{M_0 + R(L - x)}{P} + A\cos \lambda x + B \sin \lambda x, \quad \lambda^2 = \frac{P}{EI}. \tag{6.59}$$

Using the boundary conditions, $v(0) = 0$, and $v'(0) = 0$, we get

$$A = -\frac{M_0 + RL}{P}, \quad B = \frac{R}{P\lambda}. \tag{6.60}$$

Then

$$v(x) = \frac{M_0 + RL}{P}[1 - \cos \lambda x] + \frac{R}{P\lambda}[\sin \lambda x - \lambda x]. \tag{6.61}$$

The remaining conditions, $v(L) = 0$, and $v'(L) = 0$, give

$$\frac{M_0 + RL}{P}[1 - \cos \lambda L] + \frac{R}{P\lambda}[\sin \lambda L - \lambda L] = 0, \tag{6.62}$$

$$\frac{M_0 + RL}{P}\lambda \sin \lambda L + \frac{R}{P\lambda}[\lambda \cos \lambda L - \lambda] = 0. \tag{6.63}$$

Treating $(M_0 + RL)/P$ and $R/(P\lambda)$ as two unknowns, the determinant of the system gives

$$(1 - \cos \lambda L)^2 + \sin \lambda L(\sin \lambda L - \lambda L) = 0. \tag{6.64}$$

Expanding the first term, we get

$$1 - 2\cos \lambda L + \cos^2 \lambda L + \sin^2 \lambda L - \lambda L \sin \lambda L = 0, \tag{6.65}$$

which simplifies to

$$1 - \cos \lambda L - (\lambda L/2) \sin \lambda L = 0. \tag{6.66}$$

Using the identities

$$1 - \cos \lambda L = 2\sin^2(\lambda L/2), \quad \sin \lambda L = 2\cos(\lambda L/2)\sin(\lambda L/2), \tag{6.67}$$

we see that the determinant becomes,

$$2\sin(\lambda L/2)[\sin(\lambda L/2) - (\lambda L/2)\cos(\lambda L/2)] = 0. \tag{6.68}$$

Then

$$\lambda L = 2n\pi, \quad n = 1, 2, \ldots, \tag{6.69}$$

makes the first factor zero and the critical load for this solution is

$$P_{cr} = \frac{4\pi^2 EI}{L^2}, \tag{6.70}$$

which is four times the critical load for a simply supported beam. The mode shape for these eigenvalues are obtained by noting from Eq. (6.62) that $R = 0$, and the deflection has the form

$$v(x) = D(1 - \cos 2n\pi x/L). \tag{6.71}$$

From the second factor in Eq. (6.68), we find

$$\tan(\lambda L/2) = (\lambda L/2). \tag{6.72}$$

This is the characteristic equation we found for a propped cantilever beam, with L replaced by $L/2$. This equation gives a series of eigenvalues located in between the eigenvalues $2n\pi$. From Eq. (6.62), we find

$$\frac{R}{P\lambda} = \frac{M_0 + RL}{P}\frac{2}{\lambda L}, \tag{6.73}$$

and the mode shapes are given by

$$v(x) = D\left[1 - 2\frac{x}{L} - \cos\lambda x + 2\frac{\sin\lambda x}{\lambda L}\right]. \tag{6.74}$$

These are antisymmetric about the midpoint of the beam, resembling two propped cantilevers pinned at the midpoint.

6.6.1 Effective Lengths

To include the different boundary conditions and their effect on the critical load, we may define an effective Euler beam (pinned-pinned) equivalent to any special case of support conditions.

Let ℓ denote the effective length. Then for an Euler beam, by definition,

$$\ell = L. \tag{6.75}$$

For a cantilever beam

$$\ell = L/2. \tag{6.76}$$

For a propped cantilever

$$\ell = 0.7L, \tag{6.77}$$

and for a clamped-clamped beam

$$\ell = 2L. \tag{6.78}$$

If we sketch the mode shape, the effective lengths can be visualized.

Beams in a vertical position carrying compressive load are often referred to as columns.

6.7 Energy Method for Buckling of Beams

Energy method provides a means for approximating the critical loads in buckling. When the beam buckles, the applied compressive load moves. First we want to obtain the potential energy involved in this movement. We neglect the shortening of the beam due to compressive stress and attribute the shortening to bending. As shown in Fig. 6.11, a small triangle at x consisting of the sides Δx, Δv, and Δs

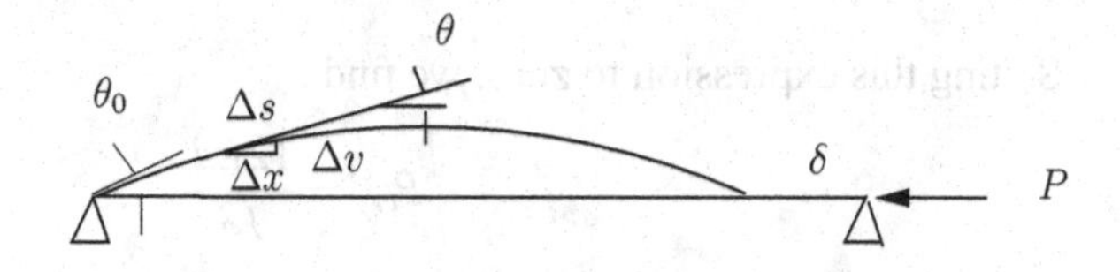

Figure 6.11 The distance moved by the compressive load during buckling.

shows

$$\Delta s^2 = \Delta x^2 + \Delta v^2. \tag{6.79}$$

Assuming the curved length of the beam along s is L,

$$\frac{dx}{ds} = \sqrt{1 - \left(\frac{dv}{ds}\right)^2}. \tag{6.80}$$

The shortening, δ, can be expressed as

$$\delta = \int_0^L \left(1 - \frac{dx}{ds}\right) ds = \int_0^L \left[1 - \sqrt{1 - \left(\frac{dv}{ds}\right)^2}\right] ds. \tag{6.81}$$

Assuming dv/ds to be small, we use the binomial theorem to expand the integrand, to arrive at

$$\delta = \frac{1}{2}\int_0^L \left(\frac{dv}{ds}\right)^2 ds. \tag{6.82}$$

With this, the potential energy of the force P_{cr} is

$$V = -P_{cr}\delta = -\frac{1}{2}P_{cr}\int_0^L \left(\frac{dv}{ds}\right)^2 ds. \tag{6.83}$$

Including the strain energy of bending, the total potential energy is

$$\Pi = \frac{1}{2}\int_0^L \left[EI(v'')^2 - P_{cr}(v')^2\right] ds, \tag{6.84}$$

where the primes indicate differentiation with s. The difference between using s and x as the variable is negligible under the assumption dv/ds is small. The true buckling mode makes the total potential energy a minimum. Any other mode shape that satisfies the displacement boundary conditions will give a higher value for Π. For example, for Euler buckling, instead of the true mode

$$v(x) = D\sin \pi x/L, \quad v(s) = D\sin \pi s/L, \tag{6.85}$$

if we use an approximate mode

$$v(s) = Ds(L - s), \tag{6.86}$$

we have

$$v'' = -2D, \quad v' = D(L - 2s). \tag{6.87}$$

$$\Pi = \frac{1}{2}\int_0^L [EI(2D)^2 - P_{cr}D^2(L - 2s)^2]ds = 2D^2L[EI - P_{cr}L^2/12]. \tag{6.88}$$

Setting this expression to zero, we find

$$P_{cr} = \frac{12EI}{L^2}. \tag{6.89}$$

This approximate critical buckling load is above the exact value of $\pi^2 EI/L^2$. We may improve the approximation using functions with more degrees of freedom.

6.8 Post-buckling Deflections

When the applied load is above the critical load, our small deflection theory predicts indefinite maximum deflections. However, we have to use large deflection theory with the exact expression for the curvature of the beam to proceed with the post-buckling calculations. The exact curvature is given by

$$\kappa = \frac{d\theta}{ds}. \tag{6.90}$$

The geometry of the deflected beam is shown in Fig. 6.11. The tangent to the neutral axis makes an angle θ with the horizontal at an arbitrary point s. This angle at $s = 0$ is denoted by θ_0. We also have

$$\frac{dv}{ds} = \sin\theta, \quad \frac{dx}{ds} = \cos\theta. \tag{6.91}$$

The equilibrium of the beam under an applied compressive load P requires

$$EI\frac{d\theta}{ds} + Pv = 0. \tag{6.92}$$

Here, when v is positive, the curvature is negative. At this point, it is advantageous to introduce nondimensional quantities

$$s \to s/L, \quad x \to x/L, \quad v \to v/L, \quad \mu^2 = PL^2/(EI), \tag{6.93}$$

and rewrite the equilibrium equation, (6.92), in the form

$$\frac{d\theta}{ds} + \mu^2 v = 0. \tag{6.94}$$

Differentiating this equation with respect to s and using $dv/ds = \sin\theta$, we get

$$\frac{d^2\theta}{ds^2} + \mu^2 \sin\theta = 0. \tag{6.95}$$

We can convert this to a first-order differential equation by multiplying by $d\theta/ds$ and integrating.

$$\frac{d\theta}{ds}\frac{d^2\theta}{ds^2} + \mu^2 \sin\theta \frac{d\theta}{ds} = 0, \tag{6.96}$$

$$\frac{1}{2}\left(\frac{d\theta}{ds}\right)^2 - \mu^2 \cos\theta = C, \tag{6.97}$$

where C is a constant of integration. From Eq. (6.92), at $s = 0$, we have $v = 0$ and $d\theta/ds = 0$. Then

$$C = -\mu^2 \cos\theta_0. \tag{6.98}$$

This leads to

$$\frac{d\theta}{ds} = -\sqrt{2}\mu\sqrt{\cos\theta - \cos\theta_0}, \tag{6.99}$$

where we have taken the negative square root to get negative curvature. We rearrange this equation to separate the variables in the form

$$\sqrt{2}\mu ds = -\frac{d\theta}{\sqrt{\cos\theta - \cos\theta_0}}. \tag{6.100}$$

Similarly, from $dx/ds = \cos\theta$ and $dv/ds = \sin\theta$, we get

$$\sqrt{2}\mu dx = -\frac{\cos\theta d\theta}{\sqrt{\cos\theta - \cos\theta_0}}, \tag{6.101}$$

$$\sqrt{2}\mu dv = -\frac{\sin\theta d\theta}{\sqrt{\cos\theta - \cos\theta_0}}. \tag{6.102}$$

Integrals of the preceding expressions are:

$$\sqrt{2}\mu s = \int_{\theta}^{\theta_0} \frac{d\theta}{\sqrt{\cos\theta - \cos\theta_0}}, \tag{6.103}$$

$$\sqrt{2}\mu x = \int_{\theta}^{\theta_0} \frac{\cos\theta d\theta}{\sqrt{\cos\theta - \cos\theta_0}}, \tag{6.104}$$

$$\sqrt{2}\mu v = \int_{\theta}^{\theta_0} \frac{\sin\theta d\theta}{\sqrt{\cos\theta - \cos\theta_0}}. \tag{6.105}$$

At this juncture, we make a few observations. The first integral gives us an equation for μ as a function of θ_0, if we assume the deflection is symmetric with respect to the point $s = 1/2$ and $\theta = 0$ at this point. Then,

$$\frac{\mu}{\sqrt{2}} = \int_{0}^{\theta_0} \frac{d\theta}{\sqrt{\cos\theta - \cos\theta_0}}. \tag{6.106}$$

The second integral can be written as

$$\sqrt{2}\mu x = \int_{\theta}^{\theta_0} \sqrt{\cos\theta - \cos\theta_0} d\theta + \sqrt{2}\mu s \cos\theta_0. \tag{6.107}$$

The third integral can be explicitly integrated to get

$$\frac{\mu}{\sqrt{2}} v = \sqrt{\cos\theta - \cos\theta_0}. \tag{6.108}$$

Traditionally, Eq. (6.106) is expressed using the Elliptic integrals, which are tabulated in mathematical handbooks. However, with our currently available software, we may do numerical evaluation of the preceding integrals. One difficulty with the integral in Eq. (6.106) is that the integrand is unbounded at $\theta = \theta_0$, that is, the integral is singular.

Near $\theta = \theta_0$, we let

$$\theta_0 - \theta = \phi, \tag{6.109}$$

which is small, and

$$\cos\theta = \cos(\theta - \theta_0 + \theta_0) = \cos\phi\cos\theta_0 + \sin\phi\sin\theta_0, \tag{6.110}$$

$$\cos\theta - \cos\theta_0 = \sin\phi\sin\theta_0 + (\cos\phi - 1)\cos\theta_0, \tag{6.111}$$

$$= \phi\sin\theta_0, \tag{6.112}$$

Table 6.1. Numerical solutions for P/P_{cr}, maximum deflection $v(1/2)$, and the location of the right end of the beam $2x(1/2)$ for various values of the angle θ_0

θ_0(in degrees)	P/P_{cr}	$v(1/2)$	$2x(1/2)$
0	1	0	1
10	1.0038	0.0554	0.9984
20	1.0154	0.1097	0.9697
30	1.0351	0.1620	0.9324
40	1.0637	0.2111	0.8812
50	1.1020	0.2563	0.8170
60	1.1517	0.2966	0.7410
70	1.2147	0.3313	0.6546
80	1.2940	0.3597	0.5600
90	1.3932	0.3814	0.4569
110	1.6779	0.4026	0.2372
130	2.1604	0.3925	0.0082
150	3.1054	0.3490	−0.2223
170	5.9505	0.2600	−0.4714
179	15.2183	0.1632	−0.6735

where we have neglected terms of the order of ϕ^2. Splitting the interval of integration $(0, \theta_0)$ into $(0, \theta_0 - \epsilon)$ and $(\theta_0 - \epsilon, \theta_0)$ where ϵ is chosen to be small, we write

$$\int_0^{\theta_0} \frac{d\theta}{\sqrt{\cos\theta - \cos\theta_0}} = I_1 + I_2,$$

$$I_1 = \int_0^{\theta_0-\epsilon} \frac{d\theta}{\sqrt{\cos\theta - \cos\theta_0}},$$

$$I_2 = \int_0^{\epsilon} \frac{d\phi}{\sqrt{\phi \sin\theta_0}},$$

$$= 2\frac{\sqrt{\epsilon}}{\sqrt{\sin\theta_0}}. \tag{6.113}$$

This way we avoid infinite integrands in our numerical integration. The quantity,

$$\mu = \sqrt{\frac{PL^2}{EI}}, \tag{6.114}$$

can be related to the ratio of the applied load P and the Euler critical load, P_{cr}, through the relation

$$\frac{P}{P_{cr}} = \frac{\mu^2}{\pi^2}. \tag{6.115}$$

As can be seen from Table 6.1, when the applied load P is increased the deflection at $s = 1/2$ first increases and then decreases. Meanwhile, the point of application of the load moves from $x = 1$ toward the left, eventually moving past the origin. Along the way, the deflected beam forms a loop. The numerical calculations were made using the software *Mathematica* by Wolfram.

6.9 Torsional-bending Buckling of a Compressed Beam

When thin-walled beams are subjected to a compressive load, they have a tendency to buckle in torsion. We will limit our discussion of this topic to thin-walled open sections with one axis (the z-axis) of symmetry. Let t denote the thickness of the section and the s-coordinate describe the mid-plane in the counter-clockwise sense. As shown in Fig 6.12, an element of length Δx and cross-sectional area

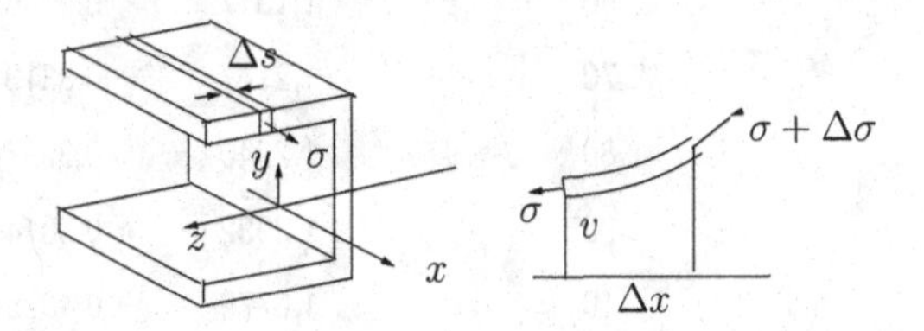

Figure 6.12 A stressed element of the beam after deformation.

$t\Delta s$ subjected to a stress σ, in the deformed shape, produces a force in the y-direction,

$$\Delta f_y \Delta x = \sigma t \Delta s[-v' + (v' + v''\Delta x)], \quad \Delta f_y = \sigma t \Delta s v''. \tag{6.116}$$

Integrating df_y over the cross-section of the beam, we get

$$f_y = v'' \int \sigma t ds = N v'', \tag{6.117}$$

where N is the net tension in the cross-section. Comparing the load-deflection relation,

$$EI\frac{d^4 v}{dx^4} = f_y, \tag{6.118}$$

and the second derivative of our buckling relation,

$$EI\frac{d^4 v}{dx^4} = -Pv'', \tag{6.119}$$

we see that a lateral loading of $(-Pv'')$ arises due to the curvature of the compressed beam.

When there is a rotation of the cross-section superposed on the displacement due to bending, the local displacement is s-dependent and so is the curvature.

In Fig. 6.13 C represents the centroid, S the shear center. Due to bending we have displacements (v, w). The displacement due to rotation about S, $\boldsymbol{u}_T$, can be expressed in vector form as $\boldsymbol{u}_T = \boldsymbol{i}\theta \times \boldsymbol{r}_s$. Using

$$\boldsymbol{r} = y\boldsymbol{j} + z\boldsymbol{k} = e\boldsymbol{k} + \boldsymbol{r}_s, \quad \boldsymbol{r}_s = y\boldsymbol{j} + (z - e)\boldsymbol{k}, \tag{6.120}$$

$$\boldsymbol{u}_T = \theta\boldsymbol{i} \times [y\boldsymbol{j} + (z - e)\boldsymbol{k}] = \theta[y\boldsymbol{k} - (z - e)\boldsymbol{j}]. \tag{6.121}$$

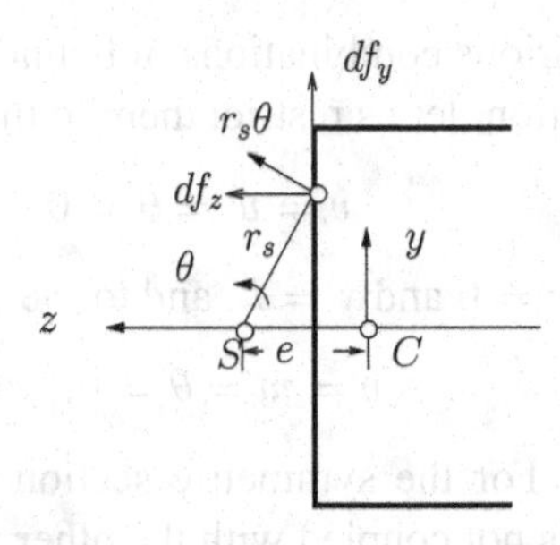

Figure 6.13 Rotation of a section about the shear center.

Then, the total displacements are $v - (z - e)\theta$ in the y-direction and $w + y\theta$ in the z-direction. On an area element tds with axial stress σ, the lateral forces are

$$df_y = \sigma tds[v'' - (z - e)\theta''], \quad df_z = \sigma tds[w'' + y\theta'']. \tag{6.122}$$

Integrating over the area using $\int ydA = \int zdA = 0$, we find

$$f_y = \sigma A[v'' + e\theta''], \quad f_z = \sigma Aw''. \tag{6.123}$$

The forces df_y and df_z also create an incremental torque per unit of length. About the shear center S, this torque due to bending, dT_B/dx, is obtained as

$$\begin{aligned} \frac{dT_B}{dx} &= -\int_A [df_z y - df_y(z - e)], \\ &= -\sigma \left[\theta'' \int_A (y^2 + z^2 + e^2)tds + Aev''\right], \\ &= -\sigma[\theta'' I_S + Aev''], \end{aligned} \tag{6.124}$$

where I_S is the polar second moment of the area about the shear center,

$$I_S = I_{yy} + I_{zz} + Ae^2 = A(\rho^2 + e^2), \tag{6.125}$$

with ρ being the radius of gyration. For a compressive applied load P,

$$\sigma = -P/A, \tag{6.126}$$

and

$$f_y = -P[v'' + e\theta''], \quad f_z = -Pw'', \quad \frac{dT_B}{dx} = P\frac{I_S}{A}\theta'' + Pev''. \tag{6.127}$$

Using the distributed forces in the deflection equations, we get

$$EI_{zz}v'''' = -P[v'' + e\theta''], \quad EI_{yy}w'''' = -Pw''. \tag{6.128}$$

From Chapter 4, Eq. (4.55), we have the torque relation

$$GJ\theta' - E\Gamma\theta''' = T_B, \tag{6.129}$$

where Γ is the Wagner torsional-bending constant for the cross section, which arises from the warping constraint. We introduce dT_B/dx, by differentiating this equation, to get

$$GJ\theta'' - E\Gamma\theta'''' = P(\rho^2 + e^2)\theta'' + Pev''. \tag{6.130}$$

Although various combinations of boundary conditions are possible to complete our formulation, let us restrict them to the simply supported case, where

$$v = w = \theta = 0, \quad v'' = w'' = \theta'', \tag{6.131}$$

at the ends $x = 0$ and $x = L$, and to the clamped-clamped case, where

$$v = w = \theta = 0, \quad v' = w' = \theta' = 0, \tag{6.132}$$

at the ends. For the symmetric section we had in mind, the deflection in the z-direction is not coupled with the other variables, and

$$P_w = \frac{\pi^2 E I_{yy}}{L^2} \tag{6.133}$$

is a critical load for the simply supported case.

6.9.1 Torsional Buckling

As a special case of the differential equation (6.130), by letting $v = 0$, we get

$$GJ\theta'' - E\Gamma\theta'''' = P(\rho^2 + e^2)\theta''. \tag{6.134}$$

We may consider two types of boundary conditions at either end:

$$\theta = 0, \quad \theta' = 0, \tag{6.135}$$

when warping is prevented and

$$\theta = 0, \quad \theta'' = 0, \tag{6.136}$$

when warping is allowed. In comparison to beams we refer to these two cases as "fixed" end or "simply supported" end, respectively. If we have simple support at both ends, using

$$\theta = C \sin \pi z/L, \tag{6.137}$$

we get the critical load

$$P_\theta = \frac{1}{\rho^2 + e^2}\left[\frac{\pi^2 E\Gamma}{L^2} + GJ\right]. \tag{6.138}$$

6.9.2 General Case

For the general case with one axis of symmetry, using the mode shapes

$$v = B \sin \pi x/L, \quad \theta = C \sin \pi x/L, \tag{6.139}$$

and the buckling loads for the uncoupled case,

$$P_v = \frac{\pi^2 E I_{zz}}{L^2}, \quad P_\theta = \frac{1}{\rho^2 + e^2}\left[\frac{\pi^2 E\Gamma}{L^2} + GJ\right], \tag{6.140}$$

the remaining differential equations give

$$(P - P_v)B + PeC = 0, \tag{6.141}$$

$$PeB + (\rho^2 + e^2)(P - P_\theta)]C = 0. \tag{6.142}$$

For a nontrivial solution of this homogeneous system, we require the determinant to vanish, that is,

$$P^2 \frac{\rho^2}{\rho^2 + e^2} - P[P_v + P_\theta] + P_v P_\theta = 0. \tag{6.143}$$

The solutions of this quadratic, P_1 and P_2, give two more critical loads. Of course, we have to use the lowest value of these critical loads as the buckling load. Note that when $e = 0$, P_v and P_θ are the solutions as the system becomes uncoupled.

6.10 Example: Split Circular Cross-section Column under Compression

A split circular cross column with radius R, thickness t, and length L is shown in Fig. 6.14. The ends of the column are built into two rigid plates, preventing

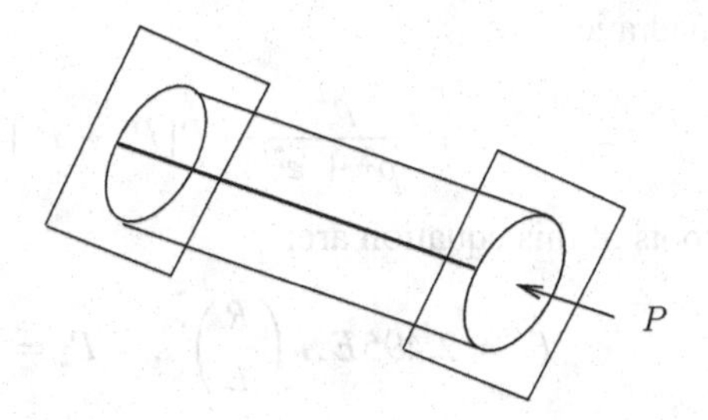

Figure 6.14 A split circular cross-section column under a load P.

warping. A compressive load P acts on the column. To simplify the calculations, we set $R^2 = tL$.

To find the buckling load we use

$$I_{yy} = I_{xx} = I = \frac{\pi R^3 t}{2}, \quad e = 2R,$$
$$J = \frac{2}{3}\pi R t^3, \quad \Gamma = \frac{2}{3} R^5 t[\pi^3 - 6\pi]. \tag{6.144}$$

For the built-in end conditions, we take

$$w = A\left[1 - \cos\frac{2\pi z}{L}\right], \quad v = B\left[1 - \cos\frac{2\pi z}{L}\right], \quad \theta = C\left[1 - \cos\frac{2\pi z}{L}\right]. \tag{6.145}$$

The differential equation for w is not coupled to the other variables and the buckling load is

$$P_w = \frac{4\pi^2 EI}{L^2} = \frac{2\pi^3 R^3 t E}{L^2} = \pi E A \left(\frac{R}{L}\right)^2, \tag{6.146}$$

where $A = 2\pi Rt$, the cross-sectional area. The constants P_v and P_θ are obtained as

$$P_v = \pi EA \left(\frac{R}{L}\right)^2, \tag{6.147}$$

$$P_\theta = \frac{1}{\rho^2 + e^2}\left[\frac{4\pi^2 E\Gamma}{L^2} + GJ\right],$$
$$= \left[4\pi(\pi^3 - 6\pi) + \frac{G}{E}\right]\frac{EA}{15}\left(\frac{R}{L}\right)^2, \tag{6.148}$$

where $\rho = R$. Using the further assumption $E/G = 2(1 + \nu) = 2.6$, we get the numerical values

$$P_v = P_w = 3.14EA\left(\frac{R}{L}\right)^2, \quad P_\theta = 10.21EA\left(\frac{R}{L}\right)^2. \tag{6.149}$$

For the critical load for the coupled bending-torsional instability we have to solve the quadratic

$$P^2\frac{\rho^2}{\rho^2 + e^2} - P[P_v + P_\theta] + P_v P_\theta = 0. \tag{6.150}$$

The roots of this equation are:

$$P_1 = 2.495EA\left(\frac{R}{L}\right)^2, \quad P_2 = 64.255EA\left(\frac{R}{L}\right)^2. \tag{6.151}$$

From this, we conclude that the minimum load for buckling is $2.495EA(R/L)^2$, with a coupled bending-torsional mode of instability.

6.11 Dynamic Stability of an Airfoil

Let us consider a simplified model of a symmetric airfoil in a steady flow of air or water with speed V. The airfoil is given two degrees of freedom: displacement v perpendicular to the flow direction and a rotation θ. Restorative forces and torques are provided by two springs. It is simpler to keep the motion v in the horizontal plane and the airfoil span vertical. This arrangement is shown in Fig. 6.15.

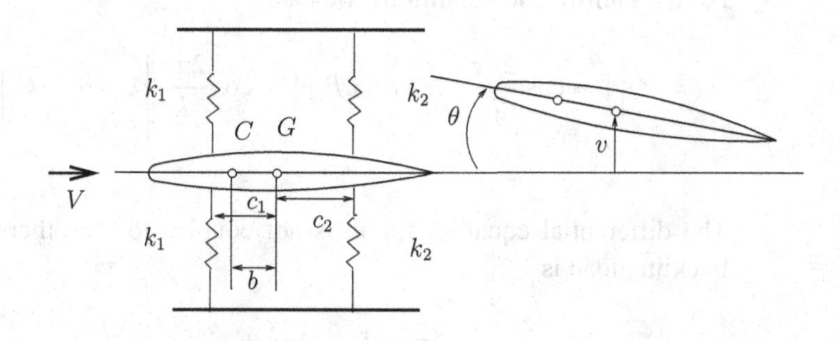

Figure 6.15 A spring-supported airfoil in a steady flow.

We may assume this setup as a long wing section suspended in a wind or water tunnel. Let C denote the aerodynamic center at quarter chord and G the center of mass.

With a lift force L acting through C, the equations of motion are:

$$m\ddot{v} = L - 2k_1(v + c_1\theta) - 2k_2(v - c_2\theta), \tag{6.152}$$

$$I\ddot{\theta} = -M + Lb - 2k_1c_1(v + c_1\theta) + 2k_2c_2(v - c_2\theta), \tag{6.153}$$

where the dots indicate time derivatives and m is the mass of the wing, I is the moment of inertia about the point G, and

$$L = \frac{1}{2}\rho V^2 S C_{L\alpha}\theta, \quad M = \frac{1}{2}\rho V^2 S c C_{M\alpha}\theta, \tag{6.154}$$

with ρ the fluid density, V the flow velocity, S the plan-form area of the wing, $C_{L\alpha}$, and $C_{M\alpha}$ the lift and moment curve slopes, and c the wing chord. We may shorten these expressions for the lift and the moment to have

$$L = L_\theta\theta, \quad L_\theta = \frac{1}{2}\rho V^2 S C_{L\alpha}. \tag{6.155}$$

$$M = M_\theta\theta, \quad M_\theta = \frac{1}{2}\rho V^2 S c C_{M\alpha}. \tag{6.156}$$

Using harmonic solutions of the form

$$v = Be^{i\omega t}, \quad \theta = Ce^{i\omega t}, \tag{6.157}$$

the equations of motion reduce to

$$\begin{bmatrix} m\omega^2 - 2(k_1 + k_2) & L_\theta - 2(k_1c_1 - k_2c_2) \\ -2(k_1c_1 - k_2c_2) & I\omega^2 + M_\theta + L_\theta b - 2(k_1c_1^2 + k_2c_2^2) \end{bmatrix} \begin{Bmatrix} B \\ C \end{Bmatrix} = \begin{Bmatrix} 0 \\ 0 \end{Bmatrix}. \tag{6.158}$$

For a nontrivial solution of this system the determinant of the matrix has to be zero. This gives a quadratic equation for the square of the angular frequency ω. If ω^2 is positive, we have oscillatory motions for v and θ. If ω^2 is negative or complex, we may have unstable solutions, which may be grouped as (a) growing in time without oscillations and (b) growing in time with oscillations. The group (a) behavior is called *divergence* and the group (b) behavior is called *flutter*.

The model we presented here is overly simplistic. In the case of a real wing attached to a plane, bending and torsional stiffnesses of the cantilevered structure replace our springs. We also have to distinguish locations of the neutral axis, shear center, center of mass, and aerodynamic center (usually the forward quarter chord). Additionally, the lift L and the moment M are not just functions of the angle of attack θ; they both depend on v, $\dot{v}$, $\ddot{v}$, $\dot{\theta}$, and $\ddot{\theta}$.

FURTHER READING

Curtis, H. D., *Fundamentals of Aircraft Structural Analysis*, Irwin (1997).
Fung, Y. C., *An Introduction to the Theory of Aeroelasticity*, Dover (1969).
Megson, T. H. G., *Aircraft Structures for Engineering Students*, Butterworth-Heinemann (2007).
Rivello, R. M., *Theory and Analysis of Flight Structures*, McGraw-Hill (1969).

Sun, C. T., *Mechanics of Aircraft Structures*, John Wiley (2006).
Timoshenko, S. P., *Theory of Elastic Stability*, McGraw-Hill (1936).

EXERCISES

6.1 Two rigid bars, each of length L, are under a compressive load of P, as shown in Fig. 6.16. There are three pin connections – two at the top and bottom supports and one in the middle. A spring of stiffness k resists the sideways deflection. Sketch the perturbed state for this configuration. Balancing the moment, obtain the critical load for the instability of this structure.

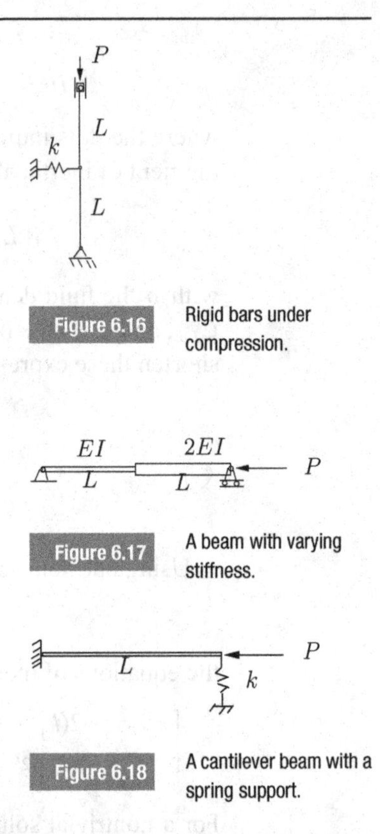

Figure 6.16 Rigid bars under compression.

Figure 6.17 A beam with varying stiffness.

Figure 6.18 A cantilever beam with a spring support.

6.2 A simply supported beam has a bending stiffness of EI on the left half of its length $2L$ and a stiffness of $2EI$ on the right half. This is shown in Fig. 6.17. Find the critical load for the buckling of this beam.

6.3 A cantilever beam with bending stiffness EI has a tip support with spring stiffness k as shown in Fig. 6.18. Find an equation for the critical load for the buckling of this beam.

6.4 For the beam and spring arrangement shown in Fig. 6.18, obtain the spring stiffness k in terms of EI and L if its critical buckling load is three times larger than that for the cantilever beam without any spring support.

6.5 A built-in beam has a bending stiffness of EI and a central support with a spring stiffness of k (see Fig. 6.19). Find an equation for the critical load for the buckling of this beam.

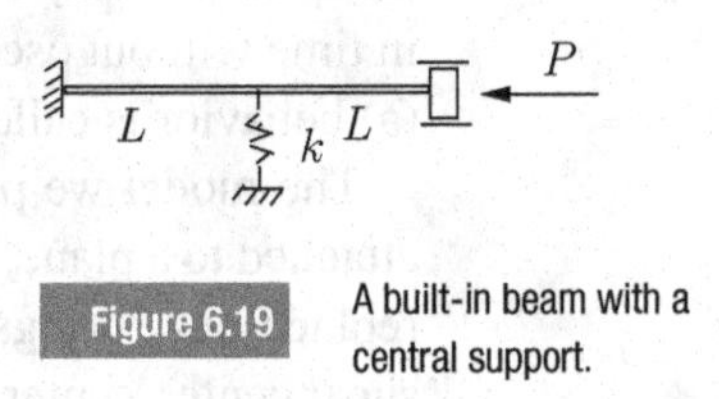

Figure 6.19 A built-in beam with a central support.

6.6 The simply supported beam in Fig. 6.20 is subjected to a compressive load P. It has a bending stiffness EI and length L. The left end of the beam has a torsional spring with spring constant $k = 2EI/L$. The function of this spring is to exert a resistive moment proportional to the slope. Obtain a transcendental equation for the constant $\mu = \lambda L$ where $\lambda^2 = P_{cr}/EI$. Find the smallest root of this equation and compare it to the Euler value.

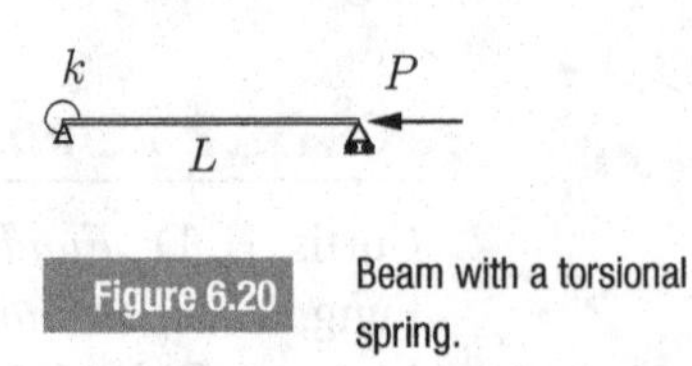

Figure 6.20 Beam with a torsional spring.

6.7 A column of length L has a symmetric cross-section (shown in Fig. 6.21) in the shape of an I, with web depth $2a$ and flange width $2a$. The web and flanges have a thickness of t. Calculate Γ, I_{yy}, I_{zz}, and J for this section. Obtain the torsional buckling load in terms of $EA(a/L)^2$, where $A = 6at$, if the ends are simply supported and warping is prevented.

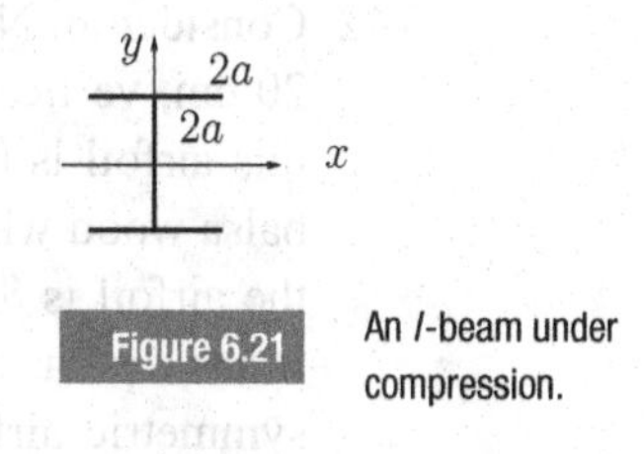

Figure 6.21 An I-beam under compression.

6.8 A simply supported column of length L has a solid circular cross-section with radius R. The column is made from a material with Young's modulus E and density ρ. If the column is spinning with an angular velocity Ω while subjected to a compressive force of P, obtain the critical value of the frequency, Ω, for instability.

6.9 A column has bending stiffness $4EI$ in its middle section of length $2a$ and EI for the first and third sections of length a. This is shown in Fig. 6.22. If it has simple supports at the ends with a compressive load P, obtain the critical load for buckling.

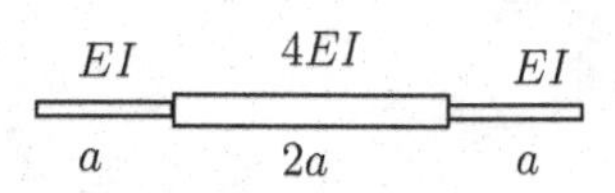

Figure 6.22 A variable stiffness column.

6.10 A column consists of two beams with bending stiffness EI pin-connected at the middle (Fig. 6.23). A torsional spring with stiffness k is attached between them. The spring exerts a moment resisting change in slope. It is free to move vertically. Obtain the critical load for this setup. Also, find its limiting cases for $k \to 0$ and $k \to \infty$.

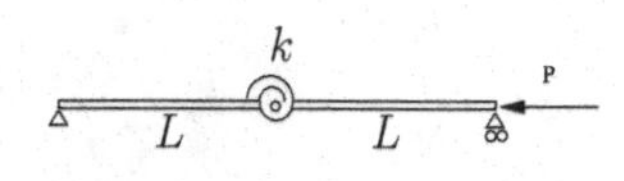

Figure 6.23 A column with a torsion spring.

6.11 The determinant of Eq. (6.158) is zero for nontrivial solutions. This can be rewritten in a simpler form, by first dividing all the terms by m and redefining B and C to obtain

$$\begin{vmatrix} \omega^2 - 2\dfrac{k_1+k_2}{m} & \dfrac{L_\theta}{mc} - 2\dfrac{k_1c_1 - k_2c_2}{mc} \\ -2\dfrac{k_1c_1 - k_2c_2}{mc} & \dfrac{I}{mc^2}\omega^2 + \dfrac{1}{mc^2}(M_\theta + L_\theta b) - 2\dfrac{k_1c_1^2 + k_2c_2^2}{mc^2} \end{vmatrix} = 0.$$

Rewrite this equation by introducing

$$\omega_1^2 = 2k_1/m, \quad \omega_2^2 = 2k_2/m, \quad \gamma_1 = c_1/c, \quad \gamma_2 = c_2/c,$$

$$\lambda = \frac{1}{2}\frac{\rho V^2 SC_{L\alpha}}{mc}, \quad \mu = \frac{I}{mc^2}, \quad \nu = \frac{b}{c}, \quad M_\theta = 0.$$

6.12 Consider an NACA0012 airfoil with a chord length of 10 cm and span of 20 cm, vertically placed in a water tunnel. The lift curve slope, $C_{L\alpha}$, for this airfoil is 0.1 per degree of angle of attack. The airfoil is made from balsa wood which has a density of 200 kg/m^3. The cross-sectional area of the airfoil is 8.17 cm^2 and its center of mass is at 4.2 cm from the nose. It has a polar moment of area 44.9 cm^4 about the center of mass. For this symmetric airfoil, you may neglect $C_{M\alpha}$ with the aerodynamic center at quarter chord. Assume both springs have $k_1 = k_2 = 10$ kN/cm. They are at $c_1 = 2.2$ cm and $c_2 = 5.0$ cm. If water has a density of 1000 kg/m^3 and the flow velocity is 10 m/s, obtain the frequency of oscillations of the airfoil.

7 Failure Theories

Under the heading of "Failure Theories" we discuss a number of topics relevant to characterizing the strength of materials. Due to their high strength-to-weight ratio and low cost, aluminum alloys have been extensively tested in the past. Steel and titanium alloys are used for high strength components at the expense of weight in the case of steel and cost in the case of titanium. All of these metals are ductile in nature and their failure is characterized by yielding and plastic deformation. The early structural material, namely, wood, and the new epoxy/fiber composite materials are characterized as brittle. In these, there is hardly any perceptible plastic deformation and they fail abruptly.

7.1 Brittle Failure

From the available experimental data, brittle failure is governed by one of the three accepted theories that follow.

7.1.1 Maximum Principal Stress Theory

On the basis of uni-axial tension and compression tests, we assign ultimate strengths σ_{ut} in tension and σ_{uc} in compression. Numerically σ_{uc} is positive. We extend these findings to two- and three-dimensional situations where three principal stresses σ_1, σ_2, and σ_3 are present. According to the maximum principal stress theory, when any one of the three principal stresses reaches ultimate stress, the material would fail. Fig. 7.1 shows the failure envelope in a two-dimensional case.

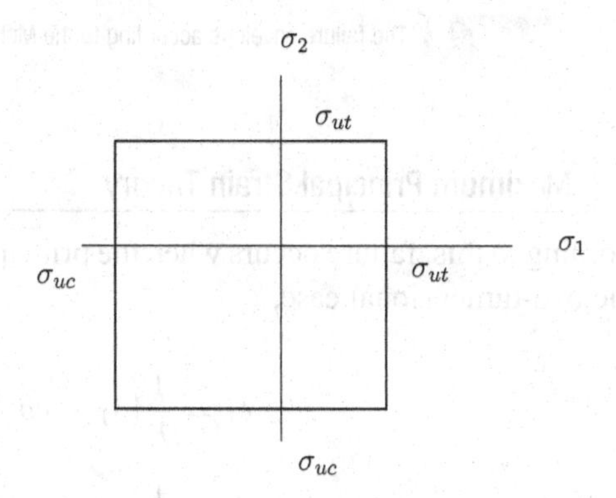

Figure 7.1 The failure envelope according to the maximum principal stress theory when $\sigma_3 = 0$.

Usually, the ultimate stress in compression is numerically higher than that in tension. Fig. 7.1 represents this unsymmetry.

7.1.2 Mohr-Coulomb Theory

The maximum principal stress theory only involves the highest absolute value of one stress. For some materials, the presence of a second stress alters the ultimate stress. Mohr-Coulomb theory is an attempt to take into account this coupling. In the case of the bi-axial stress, this theory states that failure occurs if in the

$$\text{First quadrant:} \quad \frac{\sigma_1}{\sigma_{ut}} \quad \text{or} \quad \frac{\sigma_2}{\sigma_{ut}} = 1, \tag{7.1}$$

$$\text{Third quadrant:} \quad \frac{\sigma_1}{\sigma_{uc}} \quad \text{or} \quad \frac{\sigma_2}{\sigma_{uc}} = -1, \tag{7.2}$$

$$\text{Second quadrant:} \quad \frac{\sigma_1}{\sigma_{uc}} + \frac{\sigma_2}{\sigma_{ut}} = 1, \tag{7.3}$$

$$\text{Fourth quadrant:} \quad \frac{\sigma_1}{\sigma_{ut}} + \frac{\sigma_2}{\sigma_{uc}} = 1. \tag{7.4}$$

As shown in Fig. 7.2, the effect of stress coupling is present in the second and fourth quadrants.

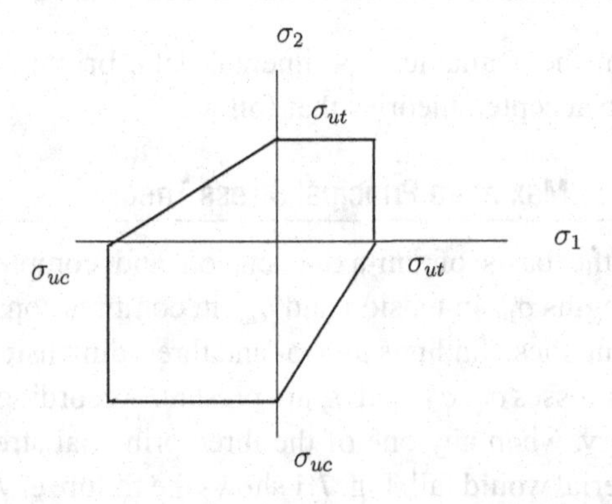

Figure 7.2 The failure envelope according to the Mohr-Coulomb theory when $\sigma_3 = 0$.

7.1.3 Maximum Principal Strain Theory

According to this, failure occurs when the principal strain reaches a critical value. In the two-dimensional case,

$$\epsilon_1 = \frac{1}{E}[\sigma_1 - \nu\sigma_2],$$

$$\epsilon_2 = \frac{1}{E}[\sigma_2 - \nu\sigma_1]. \tag{7.5}$$

Then the failure envelope in the stress plane is governed by $\sigma_1 - \nu\sigma_2 = $ a constant. This can be sketched as shown in Fig. 7.3.

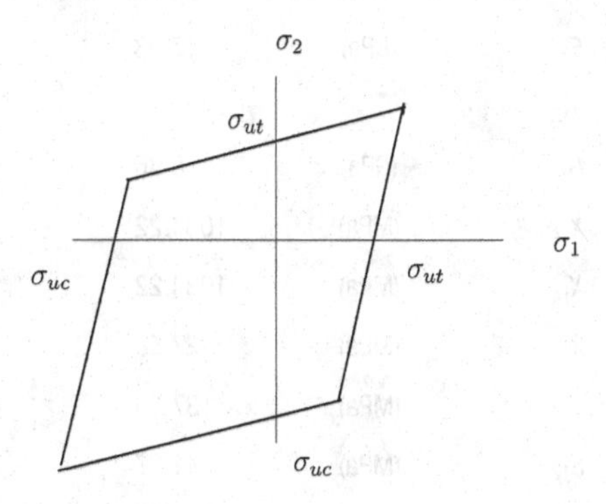

Figure 7.3 The failure envelope according to the maximum principal strain theory when $\sigma_3 = 0$.

7.2 Failure Theories for Composites

Our discussion here concerns epoxy matrix composites with glass, boron, or graphite fibers and with well-defined symmetry axes. Failure in this group of composites can be characterized as brittle. In intermetallic composites in which two or more metals form the constituents and, in general, in composites with metal components, ductile failure through yielding has to be taken into account. As shown in Fig. 7.4, we denote the fiber direction by x_1 and the mutually orthogonal directions perpendicular to the fiber directions by x_2 and x_3.

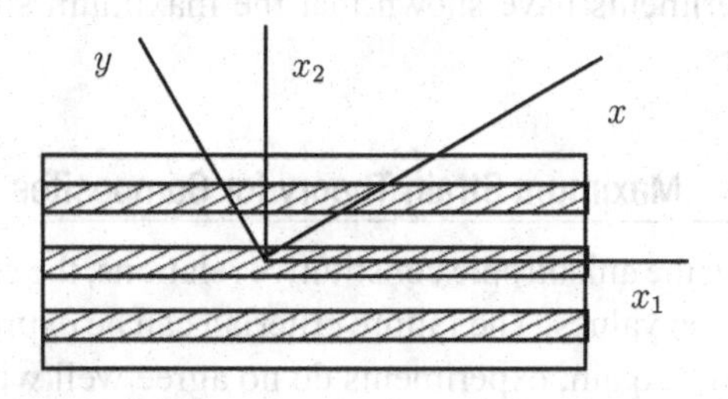

Figure 7.4 Coordinates for a fiber composite.

Uniaxial tests with a specimen oriented in the fiber direction x_1 and in the perpendicular directions x_2 and x_3 determine the failure strengths X_1, X_2, and X_3, respectively. Additional shearing tests are conducted along the planes tangential to the fibers to obtain shear strengths S_{12} and S_{13}. With these data, the objective of a failure theory is to predict failure in a multicomponent stress situation. Table 7.1 shows the elastic and failure properties of glass/epoxy, boron/epoxy, and graphite/epoxy composites. The subscripts in X_{1t} and X_{1c} indicate the values in tension and compression, respectively. We note the failure stresses in compression can be quite different from those in tension for some of these materials.

Table 7.1. Elastic and failure properties of selected composites from Jones (1975)

Property	Unit	Glass/epoxy	Boron/epoxy	Graphite/epoxy
E_1	(GPa)	53.78	206.84	206.84
E_2	(GPa)	17.93	20.68	5.17
ν_{12}		0.25	0.3	0.25
G	(GPa)	8.96	6.89	52.59
X_{1t}	(MPa)	1034.22	1379.00	1034.21
X_{1c}	(MPa)	1034.22	2757.90	689.48
X_{2t}	(MPa)	27.58	82.74	41.37
Y_{2c}	(MPa)	137.90	275.79	117.21
S_{12}	(MPa)	41.37	124.11	68.95

7.2.1 Maximum Stress Theory for Composites

If the stress components are given in a rotated coordinate system, (x, y, z), where the x-axis is at an angle θ from the x_1 axis, by using the rotation matrix $\boldsymbol{Q}$ from Chapter 2, we can find the components in the (x_1, x_2, x_3) system. The maximum stress theory predicts that to avoid failure,

$$\sigma_{11} < X_1, \quad \sigma_{22} < X_2, \quad \sigma_{33} < X_3, \quad \tau_{12} < S_{12}, \quad \tau_{13} < S_{13}. \tag{7.6}$$

In the simple case of a single stress component σ_{xx} acting in the x direction, we have the conditions

$$\sigma_{xx} < \frac{X_1}{\cos^2\theta}, \quad \sigma_{xx} < \frac{X_2}{\sin^2\theta}, \quad \sigma_{xx} < \frac{S_{12}}{\cos\theta\sin\theta}. \tag{7.7}$$

Experiments have shown that the maximum stress theory could result in large errors.

7.2.2 Maximum Strain Theory for Composites

Using the anisotropic constitutive relations, the stresses at failure can be converted to strain values. The failure criterian can be expressed as conditions on maximum strains. Again, experiments do no agree well with this approach.

7.2.3 Tsai–Hill Theory

A modified version of Hill's criterion for anisotropic plastic yield was suggested by Tsai as a failure criterion for composites. This theory has the form

$$(G+H)\sigma_{11}^2 + (H+F)\sigma_{22}^2 + (F+G)\sigma_{33}^2 - 2(H\sigma_{11}\sigma_{22} + G\sigma_{22}\sigma_{33} + F\sigma_{33}\sigma_{11}),$$
$$+2\left(L\sigma_{23}^2 + M\sigma_{13}^2 + N\sigma_{12}^2\right) = 1. \tag{7.8}$$

The three separate conditions we have seen before are now reduced to a single condition. This adds one more constant. By applying the stress components one at a time, we have, when σ_{12} and σ_{13} act alone, failure results if

$$2N = \frac{1}{S_{12}^2}, \quad 2M = \frac{1}{S_{13}^2}. \tag{7.9}$$

Using σ_{11} , σ_{22}, and σ_{33}, one at a time, we find

$$G + H = \frac{1}{X_1^2}, \quad H + F = \frac{1}{X_2^2}, \quad F + G = \frac{1}{X_3^2}. \tag{7.10}$$

We need one more shear test in the 23-direction to find

$$L = \frac{1}{S_{23}^2}. \tag{7.11}$$

If we impose the additional condition that the directions 2 and 3 are indistinguishable, we have

$$X_2 = X_3, \quad S_{12} = S_{13}. \tag{7.12}$$

This reduces the six constants to four. For the case of plane stress, this failure criterion reduces to

$$\frac{\sigma_{11}^2}{X_1^2} - \frac{\sigma_{11}\sigma_{22}}{X_1^2} + \frac{\sigma_{22}^2}{X_1^2} + \frac{\sigma_{12}^2}{S_{12}^2} = 1. \tag{7.13}$$

With a single component of stress σ_{xx}, we get

$$\frac{\cos^4\theta}{X_1^2} + \left(\frac{1}{S_{12}^2} - \frac{1}{X_1^2}\right)\cos^2\theta \sin^2\theta + \frac{\sin^4\theta}{X_2^2} = \frac{1}{\sigma_{xx}^2}. \tag{7.14}$$

This one condition replaces the three conditions of the maximum stress or strain theories. When all the normal stresses are tensile, the Tsai–Hill theory gives results close to the experimental values. In many cases, the failure strength in compression is different from that in tension. To account for this, linear terms in normal stresses are added to the Tsai–HIll criterion to obtain the Tsai-Wu tensor criterion:

$$\begin{aligned} &F_{11}\sigma_{11}^2 + F_{22}\sigma_{22}^2 + F_{33}\sigma_{33}^2 + 2(F_{12}\sigma_{11}\sigma_{22} + F_{13}\sigma_{11}\sigma_{33} + F_{31}\sigma_{33}\sigma_{11}) \\ &+ F_1\sigma_{11} + F_2\sigma_{22} + F_3\sigma_{22} = 1. \end{aligned} \tag{7.15}$$

This calls for additional compressive failure tests to determine the added three constants.

7.3 Ductile Failure

In aerospace applications, ductile materials such as metals are deemed failed if they reach plastic yield. In a uni-axial experiment some materials show a clear value of yield stress. In some, such as steels, the yield point is not clearly observable and the elastic stress corresponding to a 0.2% strain is considered the yield stress. There are two commonly used theories to extend uni-axial behavior to the three-dimensional stress state: the Tresca theory and the von Mises theory. They are both based on the assumption that hydrostatic stress has no effect on yielding.

7.3.1 Tresca Theory

The Tresca theory states that yielding begins when the maximum shear stress reaches a critical value. If the three principal stresses are σ_1, σ_2, and σ_3, the maximum shear is (see Chapter 2),

$$\tau_{max} = \text{Max}\frac{1}{2}[|\sigma_1 - \sigma_2|, \quad |\sigma_2 - \sigma_3|, \quad |\sigma_3 - \sigma_1|]. \tag{7.16}$$

Then the shear stress at yield for the one-dimensional case is

$$\tau_{max} = \tau_y = \frac{\sigma_y}{2}. \tag{7.17}$$

When $\sigma_3 = 0$, we find

$$Max.[|\sigma_1|, |\sigma_2|, |\sigma_1 - \sigma_2|] = \sigma_y. \tag{7.18}$$

A plot of these edges in the σ_1, σ_1-plane, forming the yield envelope, is shown in Fig. 7.5.

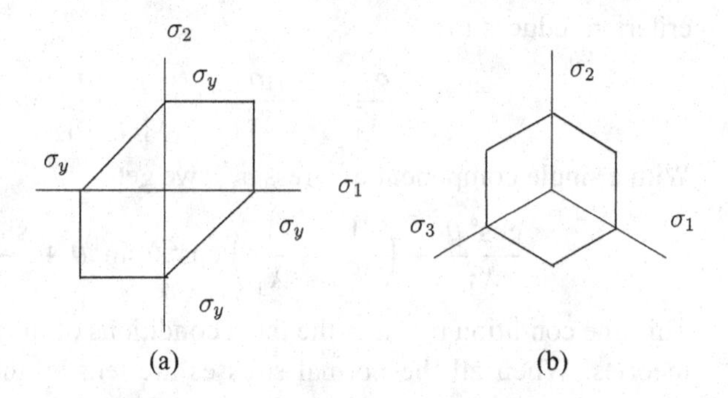

Figure 7.5 The yield envelope according to the Tresca theory when $\sigma_3 = 0$.

As the hydrostatic pressure is proportional to $\sigma_1 + \sigma_2 + \sigma_3$, the pressure independence of the yield condition implies that on any plane $\sigma_1 + \sigma_2 + \sigma_3 =$ constant, the yield envelope would appear the same. These planes are called the π-planes. If we view these planes from a perpendicular line of sight, the Tresca envelope would appear as a perfect hexagon. This is shown in Fig. 7.5(b). The corners in the hexagon may create certain problems if we need to draw normal directions to the yield surface.

7.3.2 Von Mises Theory

The yield theory of von Mises is based on the portion of the elastic energy stored in the material in the form of distortion energy. The total strain energy density is given by

$$U = \frac{1}{2}[\sigma_1\epsilon_1 + \sigma_2\epsilon_2 + \sigma_3\epsilon_3], \tag{7.19}$$

where we use the principal axes to avoid shear stresses. Using Hooke's law,

$$\epsilon_1 = \frac{1}{E}[\sigma_1 - \nu(\sigma_2 + \sigma_3)], \ldots. \tag{7.20}$$

Eliminating the strain components in terms of stress components, the total energy density becomes

$$U = \frac{1}{2E}\left[\sigma_1^2 + \sigma_2^2 + \sigma_3^2 - 2\nu(\sigma_1\sigma_2 + \sigma_2\sigma_3 + \sigma_3\sigma_1)\right]. \tag{7.21}$$

The volume change or dilatation of a unit volume is given by

$$dv = \epsilon_1 + \epsilon_2 + \epsilon_3. \tag{7.22}$$

The work done by the hydrostatic stress due to the volume change is a part of the total energy, called the dilatational energy. This is obtained as

$$\begin{aligned} U_{dil} &= \frac{1}{2}\frac{\sigma_1 + \sigma_2 + \sigma_3}{3}(\epsilon_1 + \epsilon_2 + \epsilon_3), \\ &= \frac{\sigma_1 + \sigma_2 + \sigma_3}{6}\frac{1 - 2\nu}{E}(\sigma_1 + \sigma_2 + \sigma_3), \\ &= \frac{1 - 2\nu}{6E}(\sigma_1 + \sigma_2 + \sigma_3)^2. \end{aligned} \tag{7.23}$$

If we subtract U_{dil} from U, what remains is the distortion energy U_{dis}. We may imagine a unit volume undergoing a volume change with stored energy U_{dil} and then a distortion, which stores an additional amount of energy U_{dis}.

$$\begin{aligned} U_{dis} &= U - U_{dil}, \\ &= \frac{1}{6E}\left[3(\sigma_1^2 + \sigma_2^2 + \sigma_3^2) - 6\nu(\sigma_1\sigma_2 + \sigma_2\sigma_3 + \sigma_3\sigma_1)\right] \\ &\quad - \frac{1 - 2\nu}{6E}\left[\sigma_1^2 + \sigma_2^2 + \sigma_3^2 + 2(\sigma_1\sigma_2 + \sigma_2\sigma_3 + \sigma_3\sigma_1)\right], \\ &= \frac{1 + \nu}{3E}\left[\sigma_1^2 + \sigma_2^2 + \sigma_3^2 - (\sigma_1\sigma_2 + \sigma_2\sigma_3 + \sigma_3\sigma_1)\right]. \end{aligned} \tag{7.24}$$

According to the von Mises theory, a ductile material yields when the distortion energy U_{dis} reaches a critical value. From a uni-axial test, the critical value for U_{dis} is

$$U_{dis} = \frac{1 + \nu}{3E}\sigma_y^2. \tag{7.25}$$

For a bi-axial stress state with $\sigma_3 = 0$, we have

$$\sigma_1^2 + \sigma_2^2 - \sigma_1\sigma_2 = \sigma_y^2. \tag{7.26}$$

This is an equation for an ellipse, with its major axis making 45° with the σ_1-direction. Fig. 7.6 shows the von Mises ellipse superposed on the Tresca hexagon. The ellipse becomes a circle when projected on to the π-plane. The yield behavior of steel is often modeled using either Tresca or von Mises theory.

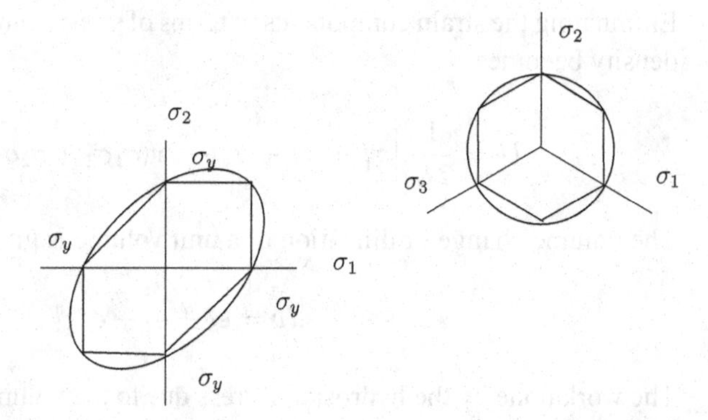

Figure 7.6 The yield envelopes according to the Tresca theory and the von Mises theory when $\sigma_3 = 0$.

7.4 Fatigue

Experiments have shown that the strength of metals under incremental loading in tension (or compression) is substantially different compared to the case of cyclic loading. Anyone who has broken a paper clip by bending it back and

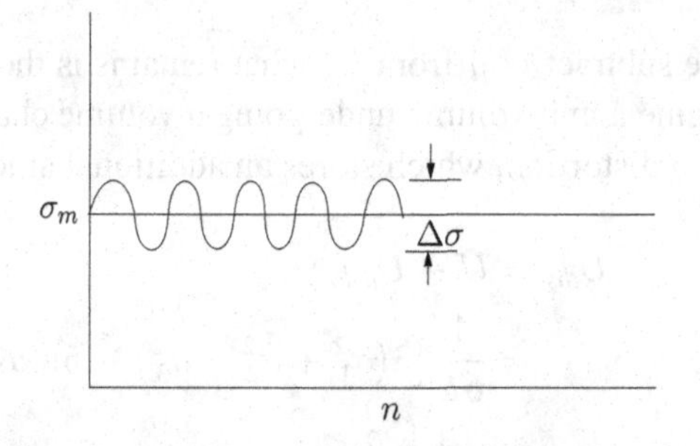

Figure 7.7 An oscillating stress with a mean value of σ_m.

forth knows this. Fig. 7.7 shows an oscillating stress amplitude around a steady mean stress of σ_m. The difference between the maximum and minimum stress of the oscillation is usually denoted by $\Delta\sigma$. First, let us consider the case of zero mean stress and the applied stress varying between $(+\Delta\sigma/2)$ and $(-\Delta\sigma/2)$. A plot, (Fig. 7.8), of the stress difference $\Delta\sigma$ versus $\log N$, called an $S-N$–plot, shows the number of load cycles N before a specimen fails for a given $\Delta\sigma$.

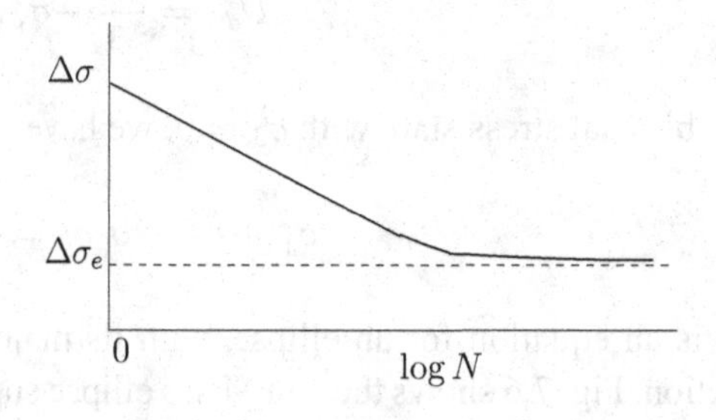

Figure 7.8 A plot of the stress amplitude versus number of cycles to failure for a typical metal.

Most metals (notably, steel) show a threshold for $\Delta\sigma$ called the endurance limit, $\Delta\sigma_e$, below which the specimen can withstand an unlimited number of cycles without failure. Above this threshold value, we see a linear range, which can be expressed as

$$\Delta\sigma = \Delta\sigma_1 - k \log N, \tag{7.27}$$

where $(-k)$ is the slope of the $S - N$ curve and $\Delta\sigma_1$ is twice the failure stress in noncyclic loading.

Typical aluminum alloys do not show an endurance limit.

7.4.1 Palmgren-Minor Rule of Cumulative Damage

Assuming that a specimen under a stress difference of $\Delta\sigma_i$ has a life of N_i cycles, if it is subjected to n_i cycles, we conclude that n_i/N_i fraction of its life has been used up. This is the basis of the Palmgren-Minor rule of cumulative damage. If we subject a structure to M different stresses $\Delta\sigma_i$, $i = 1, 2, \ldots, M$, each lasting for n_i cycles, then the Palmgren-Minor rule states,

$$\frac{n_1}{N_1} + \frac{n_2}{N_2} + \cdots + \frac{n_M}{N_M} = 1. \tag{7.28}$$

This approximate rule can be of use in predicting the life of a structure.

7.4.2 Goodman Diagram

So far we have assumed a purely oscillating stress with no mean value. When there is a positive mean value σ_m, the maximum tensile stress in the specimen is $\sigma_m + 0.5\Delta\sigma$ and this high tensile stress has a damaging effect and decreases

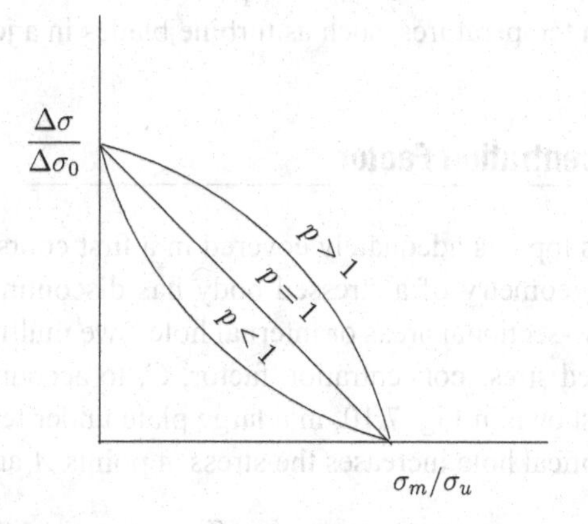

Figure 7.9 The effect of mean stress on the effective stress amplitude according to the Goodman formula.

the fatigue life of a specimen. If σ_u is the ultimate stress for failure, the Goodman diagram, shown in Fig. 7.9, shows

$$\frac{\Delta\sigma}{\Delta\sigma_0} = 1 - \frac{\sigma_m}{\sigma_u}, \tag{7.29}$$

where $\Delta\sigma$ is the effective stress difference to be used in the $S-N$ curve. This equation was modified to include the variability in the experimental data, to read

$$\frac{\Delta\sigma}{\Delta\sigma_0} = 1 - \left(\frac{\sigma_m}{\sigma_u}\right)^p, \tag{7.30}$$

where the index p is a constant that is material dependent.

7.5 Creep

Time-dependent deformation of structures under stress is called creep. Creep is particularly significant at elevated temperatures in metals. Although Hooke's law gives an instantaneous response to the applied stress, when creep is present the structure continues to deform as a function of time. Immediately after applying the load the strain increases in the form

$$\frac{d\epsilon}{dt} = Ct^{\beta}\sigma^{m}, \tag{7.31}$$

which is called primary creep. Here, C, β, and m are material constants and t is time. In the second stage, we see the strain increase in the form

$$\frac{d\epsilon}{dt} = C\sigma^{n}, \tag{7.32}$$

which is called the secondary creep. In the third stage, the strain, again, rapidly increases resulting in material failure. This stage is called the tertiary creep. All of the constants listed in the creep formulas are temperature dependent. Accurate calculation of creep deformation is crucial in applications involving high temperatures, such as turbine blades in a jet engine.

7.6 Stress Concentration Factor

This topic is adequately covered in a first course in strength of materials. When the geometry of a stressed body has discontinuities such as abrupt changes in cross-sectional areas or internal holes, we multiply the nominal stress by a factor called stress concentration factor, C, to account for the actual stress at a point. As shown in Fig. 7.10, in a large plate under tension σ_0, the presence of a small elliptical hole increases the stress at points A and B by

$$\frac{\sigma}{\sigma_0} = C = 1 + 2\frac{a}{b}, \tag{7.33}$$

where a and b are semi-axes of the ellipse.

When $b \to 0$, the stress concentration factor goes to infinity while the ellipse becomes a sharp-edged crack. Cracks are considered in the next section.

It is a recommended design practice to provide generous radii at corners to reduce stress concentration when the geometry demands such corners.

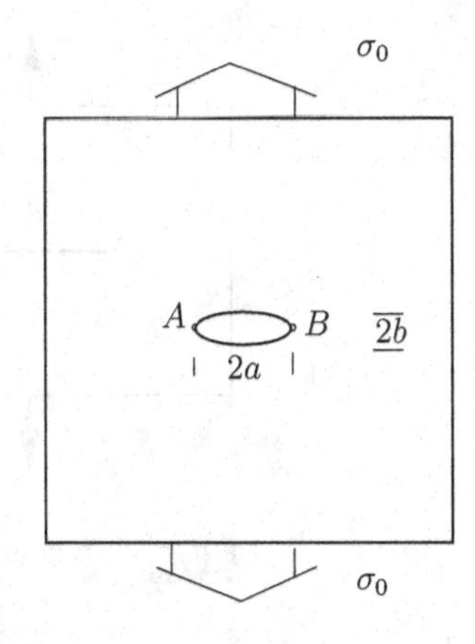

Figure 7.10 An elliptic hole in an infinite plate.

7.7 Fracture Mechanics

Materials tend to develop sharp-edged discontinuities called cracks for various reasons. Local material failure, corrosion, and dislocation accumulation in polycrystalline metals are some of the reasons for the origin of cracks. When a cracked material is stressed, one or more stress components at the tip of a crack can reach very high values – in fact, they reach infinity for a sharp crack tip according to the linear elasticity theory. The loading at a crack tip is grouped into three modes: Mode I, Mode II, and Mode III. Fig. 7.11 shows the three modes of loading in a split beam.

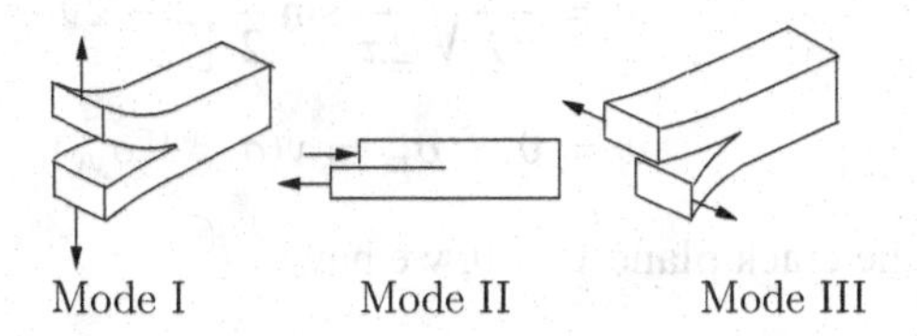

Figure 7.11 The three modes of loading affecting the crack tip stresses.

Of course, every crack has two surfaces, like the upper and lower surfaces shown in the figure. In Mode I, the displacements on these surfaces are perpendicular to the surfaces and in Mode II and Mode III they are parallel to the surfaces. As stated earlier, according to the linear elasticity solutions, some stress components at the tip of a crack may go to infinity. Using a polar coordinate system with its origin at the crack tip (see Fig. 7.12), the stress components and displacements very close to the crack tip (when r is small compared to the crack length) have been found using the theory of elasticity. These are listed in the following sections for the three modes of loading. The proportionality constants in these expressions, K_I, K_{II}, and K_{III}, are called the stress intensity factors, which depend on the shape of the body, length of the crack, and the nature of loading. When there is a single load P as shown in Fig. 7.11, the stress intensity factors are proportional to P.

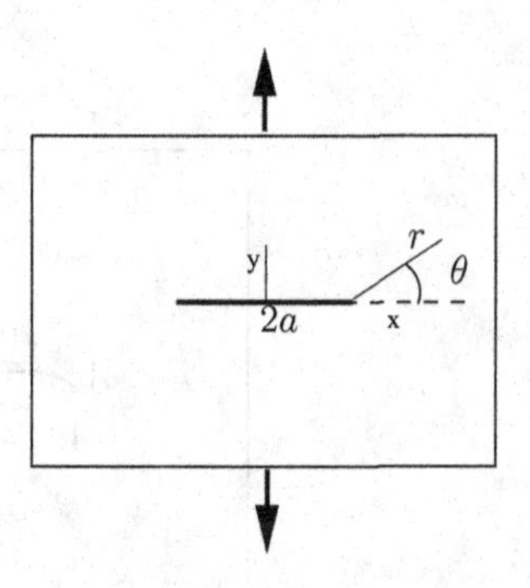

Figure 7.12 Crack tip coordinates.

7.7.1 Mode I

$$\sigma_{xx} = \frac{K_I}{\sqrt{2\pi r}} \cos\frac{\theta}{2}\left[1 - \sin\frac{\theta}{2}\sin\frac{3\theta}{2}\right],$$

$$\sigma_{yy} = \frac{K_I}{\sqrt{2\pi r}} \cos\frac{\theta}{2}\left[1 + \sin\frac{\theta}{2}\sin\frac{3\theta}{2}\right],$$

$$\sigma_{xy} = \frac{K_I}{\sqrt{2\pi r}} \sin\frac{\theta}{2}\cos\frac{\theta}{2}\cos\frac{3\theta}{2},$$

$$u = \frac{K_I}{G}\sqrt{\frac{r}{2\pi}} \cos\frac{\theta}{2}\left[1 - 2\nu + \sin^2\frac{\theta}{2}\right],$$

$$v = \frac{K_I}{G}\sqrt{\frac{r}{2\pi}} \sin\frac{\theta}{2}\left[2 - 2\nu - \cos^2\frac{\theta}{2}\right],$$

$$w = 0, \quad \sigma_{zz} = \nu(\sigma_{xx} + \sigma_{yy}), \quad \sigma_{xz} = \sigma_{yz} = 0. \tag{7.34}$$

On the crack plane $y = 0$, we have

$$\sigma_{yy} = \frac{K_I}{\sqrt{2\pi r}}, \quad x > a, \tag{7.35}$$

$$v = \frac{2(1-\nu)K_I}{G}\sqrt{\frac{r}{2\pi}}, \quad x < a. \tag{7.36}$$

An infinite panel with uniform tensile stress σ_0 at infinity with a central crack of length $2a$ is shown in Fig. 7.13. For this geometry and loading, we have

$$K_I = \sigma_0\sqrt{\pi a}. \tag{7.37}$$

For a panel with finite width W, with the crack symmetrically located, the approximation

$$K_I = \sigma_0\sqrt{\pi a}\sqrt{\sec(\pi a/W)} \tag{7.38}$$

is used.

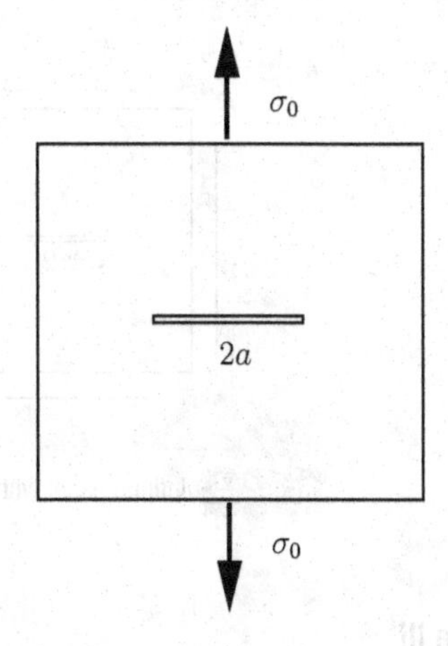

Figure 7.13 Infinite panel with a central crack under tension.

7.7.2 Mode II

$$\sigma_{xx} = -\frac{K_{II}}{\sqrt{2\pi r}} \sin\frac{\theta}{2}\left[2 + \cos\frac{\theta}{2}\cos\frac{3\theta}{2}\right],$$

$$\sigma_{yy} = \frac{K_{II}}{\sqrt{2\pi r}} \sin\frac{\theta}{2}\cos\frac{\theta}{2}\cos\frac{3\theta}{2},$$

$$\sigma_{xy} = \frac{K_{II}}{\sqrt{2\pi r}} \cos\frac{\theta}{2}\left[1 - \sin\frac{\theta}{2}\sin\frac{3\theta}{2}\right],$$

$$u = \frac{K_{II}}{G}\sqrt{\frac{r}{2\pi}} \sin\frac{\theta}{2}\left[2 - 2\nu + \cos^2\frac{\theta}{2}\right],$$

$$v = \frac{K_{II}}{G}\sqrt{\frac{r}{2\pi}} \cos\frac{\theta}{2}\left[-1 + 2\nu + \sin^2\frac{\theta}{2}\right],$$

$$w = 0, \quad \sigma_{zz} = \nu(\sigma_{xx} + \sigma_{yy}), \quad \sigma_{xz} = \sigma_{yz} = 0. \tag{7.39}$$

On the crack plane $y = 0$, we have

$$\sigma_{xy} = \frac{K_{II}}{\sqrt{2\pi r}}, \quad x > a, \tag{7.40}$$

$$u = \frac{2(1-\nu)K_{II}}{G}\sqrt{\frac{r}{2\pi}}, \quad x < a. \tag{7.41}$$

For an infinite plate with a central crack of length $2a$ (shown in Fig. 7.14) under a shear stress τ_0, we have

$$K_{II} = \tau_0\sqrt{\pi a}. \tag{7.42}$$

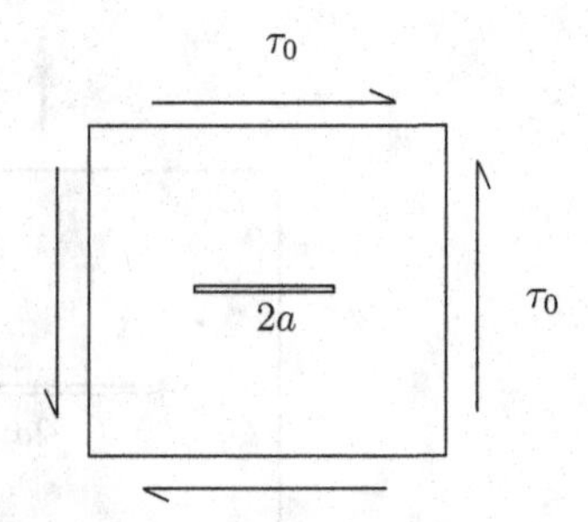

Figure 7.14 Infinite panel with a central crack under shear.

7.7.3 Mode III

$$\sigma_{xz} = -\frac{K_{III}}{\sqrt{2\pi r}} \sin\frac{\theta}{2},$$

$$\sigma_{yz} = \frac{K_{III}}{\sqrt{2\pi r}} \cos\frac{\theta}{2},$$

$$w = \frac{K_{III}}{G}\sqrt{\frac{r}{2\pi}} \sin\frac{\theta}{2},$$

$$\sigma_{xy} = 0, \quad \sigma_{xx} = \sigma_{yy} = \sigma_{zz} = 0,$$

$$u = 0 \quad v = 0, \quad \sigma_{xy} = 0. \tag{7.43}$$

On the crack plane $y = 0$, we have

$$\sigma_{yz} = \frac{K_{III}}{\sqrt{2\pi r}}, \quad \sigma_{xz} = 0, \quad x > a, \tag{7.44}$$

$$w = \frac{K_{III}}{G}\sqrt{\frac{r}{2\pi}}, \quad x < a. \tag{7.45}$$

All of the preceding expressions for Mode I and Mode II are applicable to the case of plane strain. By changing Poisson's ratio ν to $\nu/(1+\nu)$, they can be converted for use in plane stress cases.

7.7.4 Crack Propagation

When a crack propagates in a brittle material, new surfaces are created. A certain amount of energy is needed to create new surfaces. We use γ to denote the surface energy density per unit area of the surface, which is dependent on the material. For simplicity, let us consider a material with a crack of length a with an applied load P and the displacement under the load $\bar{u}$. Next, let us imagine the crack extending by Δa. The change in potential energy is given by

$$\Delta U - P\Delta\bar{u} - T\Delta S + 2\gamma t\Delta a = 0, \tag{7.46}$$

where U is the strain energy, T temperature, S entropy, and t crack depth. If we neglect the temperature effect, we can drop the entropy term. The strain energy U depends on the displacement $\bar{u}$ and on the crack length a. Thus,

$$\frac{\partial U}{\partial \bar{u}}\Delta\bar{u} + \frac{\partial U}{\partial a}\Delta a - P\Delta\bar{u} + 2\gamma\Delta a = 0. \tag{7.47}$$

Setting the coefficients of $\Delta\bar{u}$ and Δa to zero, we have

$$\frac{\partial U}{\partial \bar{u}} = P, \quad \frac{\partial U}{\partial a} = -2\gamma t, \tag{7.48}$$

where the first relation is Castigliano's second theorem and the second is a condition to propagate the crack.

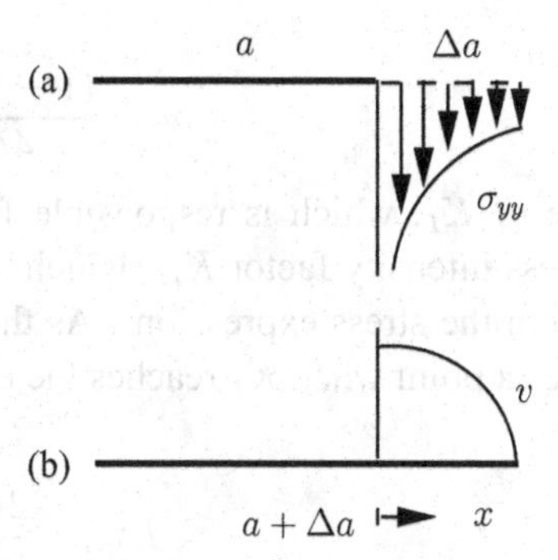

Figure 7.15 Stress in front of a crack of length a (a) and the displacement inside a crack of length $a + \Delta a$ (b).

Now, to find the rate of change in strain energy with respect to the crack length, we consider as an example Mode I loading and the state of stress in front of the original crack of length a and the displacement between $x = a$ and $x = a + \Delta a$. These are sketched in Fig. 7.15. If we isolate the upper half above the crack and use a coordinate x starting from the old crack tip, before we had the tensile stress distribution

$$\sigma_{yy} = \frac{K_I}{\sqrt{2\pi x}}, \quad x > 0, \tag{7.49}$$

between $x = 0$ and $x = \Delta a$. After the crack has extended, this stress has reduced to zero and a displacement distribution,

$$v = \frac{2(1-\nu)K_I}{G}\sqrt{\frac{\Delta a - x}{2\pi}}, \tag{7.50}$$

represents the opened crack. The transition from "before" to "after" can be viewed as decreasing the tensile stress to zero while this stress moves by the displacement v. The work done on the upper and lower half of the material is

$$\Delta U = -2\int_0^{\Delta a} \frac{1}{2}\sigma_{yy} v t dx,$$

$$= -\frac{(1-\nu)K_I^2}{2\pi G}\int_0^{\Delta a}\frac{\sqrt{\Delta a - x}}{\sqrt{x}} t dx,$$

$$\frac{\partial U}{\partial a} = -\frac{(1-\nu)K_I^2}{2\pi G} t \int_0^1 \sqrt{\frac{1-x}{x}} dx, \tag{7.51}$$

where we have replaced $x/\Delta a$ by x in the last line. Using $x = \sin^2\theta$,

$$\int_0^1 \sqrt{\frac{1-x}{x}} dx = 2\int_0^{\pi/2} \cos^2\theta d\theta = \pi. \tag{7.52}$$

Finally,

$$\gamma = \frac{(1-\nu)K_I^2}{4G}. \tag{7.53}$$

This value of K_I, which is responsible for extending the crack, is called the critical stress intensity factor K_{Ic}, which has the dimension of Pa$\sqrt{m}$ as can be verified from the stress expressions. As the applied load P is increased from 0, there comes a point when K_I reaches the critical value K_{Ic} and the crack begins to propagate.

$$K_{Ic}^2 = \frac{2G\gamma}{1-\nu}. \tag{7.54}$$

Using the energy method we also get

$$K_{IIc}^2 = \frac{4G\gamma}{1-\nu}, \quad K_{IIIc}^2 = 4G\gamma. \tag{7.55}$$

When the loading is such that all the three modes are present, by computing

$$\int_0^{\Delta a} (\sigma_{yy} v + \sigma_{xy} u + \sigma_{yz} w) t dx, \tag{7.56}$$

we find

$$(1-\nu)(K_I^2 + K_{II}^2) + K_{III}^2 = 4G\gamma. \tag{7.57}$$

The total energy dissipated by creating new surfaces is also known as the energy release rate. They are denoted by G_I, G_{II}, and G_{III} for the three modes. By multiplying $2\gamma t$ by the number of active (where the crack tip is moving) crack tips, we can find the energy release rate. The energy used for creating the new surfaces can be regained, in principle, by exerting pressure and by waiting for diffusive healing of the crack in time. For all practical purposes, the term "dissipated energy" aptly describes the energy spent in creating the surfaces during the crack tip motion.

7.7.5 Example: Energy Method for a Split Cantilever

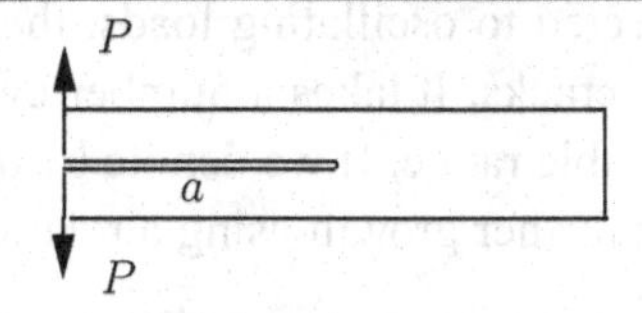

Figure 7.16 Split cantilever.

Fig. 7.16 shows a split cantilever beam with Mode I loading. For the loading shown, the deflection $\bar{u}$ of the tip is given by

$$\bar{u} = \frac{Pa^3}{3EI}, \quad P = \frac{3EI\bar{u}}{a^3}, \tag{7.58}$$

where I is the second moment of area for one of the beams. The energy stored in the upper and lower beams totals to

$$U = P\bar{u} = \frac{3EI\bar{u}^2}{a^3}. \tag{7.59}$$

Using the energy release rate

$$2\gamma t = \frac{2(1-\nu)tK_I^2}{G} = -\frac{\partial U}{\partial a} = \frac{9EI\bar{u}^2}{a^4} = \frac{P^2a^2}{EI}. \tag{7.60}$$

$$K_I = \frac{Pa}{\sqrt{4(1-\nu^2)tI}}. \tag{7.61}$$

7.7.6 Ductility at the Crack Tip

In ductile materials, the magnitude of stress is limited by the yield stress. Then the infinite stress predicted by the theory of elasticity does not occur. If the crack is away from any material boundaries, there is a small region near the crack tip where yielding takes place. As the crack advances, energy is dissipated in the yield zone. It is known that this plastic energy dissipation is much higher than the surface energy spent. However, experiments show that a critical value K_{Ic} can be used to characterize crack growth even when crack tip yielding is present. Dugdale has postulated that an effective crack length can be defined by adding an approximate width of the plastic zone to the physical crack length. Later, Irwin estimated the added crack length as the radius of an approximate plastic zone,

$$r_p = \frac{(1-2\nu)^2K_I^2}{2\pi\sigma_Y^2}, \tag{7.62}$$

for plane strain cracks in Mode I.

7.8 Fatigue Crack Growth

Poly-crystalline solids contain a large number of microscopic flaws called dislocations where the regular crystal arrangement is broken. They also contain grain boundaries where crystals with different orientations meet. When these materials are subjected to oscillating loads, the defects could move and coalesce into microscopic cracks. It takes a number cycles of loading for this microcracks to grow into visible range. If we denote by a_0 the minimum size detectable, we can predict their further growth using a relation advanced by Paris:

$$\frac{da}{dN} = C(\Delta K)^m, \tag{7.63}$$

where C and m are material constants and

$$\Delta K = K_{max} - K_{min}. \tag{7.64}$$

Experimental data are often presented in a log–log plot to obtain the constants. As an example, with K in the units of MPa$\sqrt{m}$ and da/dN in m/cycle, austenitic steel has $C = 5.6 \times 10^{-12}$ and $m = 3.25$.

FURTHER READING

Anderson, T. L., *Fracture Mechanics: Fundamentals and Applications*, CRC Press (1995).
Bannantine, J., Comer, J., and Handrock, J., *Fundamentals of Metal Fatigue Analysis*, Prentice Hall (1990).
Cherepanov, G. P., *Mechanics of Brittle Fracture*, McGraw-Hill (1979).
Jones, R. M., *Mechanics of Composite Materials*, McGraw-Hill (1975).
Lubliner, J., *Plasticity Theory*, Macmillan (1990).
Suresh, S., *Fatigue of Materials*, Cambridge University Press (1998).

EXERCISES

7.1 A brittle material fails at 80 MPa in a uni-axial test. A plate made of this material is subjected to a bi-axial tension in the x, y-plane with the proportional loading $\sigma_{xx} = 2\alpha$ and $\sigma_{yy} = \alpha$ with α being a load control parameter. Obtain the value of α at failure, (a) if the material fails according to the maximum stress criterion, and (b) if it fails under the maximum strain criterion. Assume the Poisson's ratio is $1/3$.

7.2 The yield stress for a steel alloy is 1.5 GPa. Obtain the internal pressure in a thin-walled cylindrical pressure vessel with radius 1 m and thickness 0.2 cm for yielding to occur if (a) Tresca yield criterion is used and (b) von Mises criterion is used.

7.3 A shaft of radius 5 cm carries a torque of 100 Nm and a bending moment of 50 Nm. Obtain the maximum shear stress and the maximum Mises stress in the shaft.

7.4 A circular cross-section fuselage is idealized as a thin-walled shell of radius 4 m with 48 booms of radius 2 cm each. The booms are distributed uniformly around with the first and the 25th booms being located the farthest from the neutral axis. If the maximum allowable compressive stress in the booms is 1 GPa, what is the allowable bending moment for this structure?

7.5 A boron/epoxy composite lamina is subjected to a uni-axial stress in a direction 30° to the fiber direction. Using the strength values given Table 7.1, obtain the maximum tensile and compressive stress it can carry.

7.6 A thin sheet is in plane stress in the x, y-plane with a central crack of length 5 mm, aligned with the x-axis. It is subjected to a constant shear stress, $\sigma_{xy} = 80$ MPa and zero normal stress σ_{xx}. It was observed that the crack begins to propagate when $\sigma_{yy} = 100$ MPa. Obtain the surface energy density γ for this material if it has a shear modulus of 80 GPa and a Poisson's ratio of 0.3. We may use the infinite plate approximation for this case.

7.7 A structure in plane strain has a crack of length 1 cm perpendicular to the direction of an applied stress of 100 MPa. If the yield stress for this material is 200 MPa and the Poisson's ratio is 0.3, compute the effective crack length including the plastic zone length (according to Irwin's approximation) with K_I based on the original crack length. Repeat this calculation using the effective crack length.

7.8 In the Paris law for crack growth under cycling loading,

$$\frac{da}{dN} = C(\Delta K)^n,$$

the material constants C and n are given by

$$C = 5.6 \times 10^{-12}, \quad n = 3.25,$$

for austenitic steel. These constants correspond to stress intensity factors in the units of MPa$\sqrt{\text{m}}$ and da/dN in m/cycle. Compute the number of cycles for a crack of initial length 2 mm to grow to 10 mm under an applied cyclic stress from 0 to 100 MPa. Assume the crack is located in an infinite plate perpendicular to the direction of stress.

7.9 In the Paris law for crack growth under cycling loading,

$$\frac{da}{dN} = C(\Delta K)^n,$$

the material constants C and n are given by

$$C = 5.6 \times 10^{-12}, \quad n = 3.25,$$

for austenitic steel. An infinite plate subjected to an applied stress cycle from 0 to 100 MPa for 100,000 cycles showed, during inspection, a crack of length 1 cm. Compute the initial crack length before the loading.

7.10 A plot of the stress amplitude S in cyclic loading for a material follows the law

$$S = 20 - k \log N,$$

where S is in kPa, log has the base 10, k is a constant, and N is the number of cycles to failure. If a specimen made of this material fails in 100,000 cycles at a stress amplitude of 100 kPa, find the value of k. A structure made of this material is first subjected to 100 kPa for 40,000 cycles and then to 50 kPa load cycles. How many cycles would it last at this second load?

Index